AF576030

The Applications of Plasma Techniques II

The Applications of Plasma Techniques II

Editor

Mariusz Jasiński

MDPI • Basel • Beijing • Wuhan • Barcelona • Belgrade • Manchester • Tokyo • Cluj • Tianjin

Editor
Mariusz Jasiński
Institute of Fluid Flow Machinery
Polish Academy of Sciences
Poland

Editorial Office
MDPI
St. Alban-Anlage 66
4052 Basel, Switzerland

This is a reprint of articles from the Special Issue published online in the open access journal *Applied Sciences* (ISSN 2076-3417) (available at: https://www.mdpi.com/journal/applsci/special_issues/Plasma_Techniques_II).

For citation purposes, cite each article independently as indicated on the article page online and as indicated below:

LastName, A.A.; LastName, B.B.; LastName, C.C. Article Title. *Journal Name* **Year**, *Volume Number*, Page Range.

ISBN 978-3-0365-3897-6 (Hbk)
ISBN 978-3-0365-3898-3 (PDF)

Contents

About the Editor

Mariusz Jasiński (Associate professor)

Mariusz Jasiński is a graduate of the Faculty of Technical Physics and Applied Mathematics at Gdańsk University of Technology. The M.Sc. degree in technical physics he achieved in 1997. Doctoral degree in technical sciences he defended in 2003 at the Institute of Fluid Flow Machinery of the Polish Academy of Sciences, where he is currently working. He received D.Sc. degree in technical sciences from the Lublin University of Technology (Faculty of Electrical Engineering and Computer Science) in 2015. On 1 January 2006, he was appointed head of the Department of Plasma Electrodynamics in the Centre for Plasma and Laser Engineering at the Institute of Fluid Flow Machinery of the Polish Academy of Sciences. In 2009, the unit has changed its name to the Department of Hydrogen Energy. In 2015 he was appointed associate professor at the Institute of Fluid Flow Machinery of the Polish Academy of Sciences. His research focused on the development of microwave plasma sources, plasma diagnostics based on spectroscopic techniques, and applications of plasma techniques in the energetics, electronics, surface engineering and environmental protection.

Preface to "The Applications of Plasma Techniques II"

As a Guest Editor of this Special Issue of "The Applications of Plasma Techniques II" in the "Optics and Lasers" section of the journal *Applied Sciences*, invited to write an Editorial, I briefly discuss the results of all articles published in this Special Issue below.

Mariusz Jasiński
Editor

Editorial

The Applications of Plasma Techniques II

Mariusz Jasiński

Institute of Fluid Flow Machinery, Polish Academy of Sciences, Fiszera 14, 80-231 Gdańsk, Poland; mj@imp.gda.pl

1. Introduction

This Special Issue "The Applications of Plasma Techniques II" in the section "Optics and Lasers" of the journal *Applied Sciences* is intended to provide a description of devices and processes related to plasma applications in the broad sense. Plasma is called the fourth state of matter because its properties differ significantly from those of gas. Plasma can be defined as a conductive medium generated by the ionization of gas. Thus, it occurs as a mixture of photons, electrons, and ions, but it can also contain neutral atoms and molecules. The concept of plasma includes media with very different properties. Densities and kinetic energies of plasma components differ for various types of plasma by several or even more orders of magnitude. Hence, plasmas can have very different applications. Nowadays, plasma is very common in everyday life—from ubiquitous discharge lamps to plasma TVs. In technology, plasma is used in areas as diverse as gas purification, production of chemical compounds, surface treatment of materials, synthesis of nanoparticles, and deactivation of bacteria, viruses, and cancer cells. Readers interested in this modern field of science and technology are invited to enjoy this collection of articles, which will certainly excite the curiosity of both scientists, engineers and medics interested in plasma applications. As a guest editor of this Special Issue I wish you a pleasant reading.

2. Results

As a guest editor of this Special Issue of "The Applications of Plasma Techniques II" in the "Optics and Lasers" section of the journal "*Applied Sciences*", invited to write an *Editorial*, below I briefly discuss the results of all articles published in this Special Issue.

2.1. Dual-Frequency Microwave Plasma Source

Chi Chen et al. in [1] presented a dual-frequency microwave plasma source based on microwave coaxial transmission line, and 915 and 2450 MHz microwaves were used in this study. Two waves were delivered from two ports into the plasma reactor. One of these waves was used to excite the plasma and the other to regulate the plasma characteristics. In this system, the electron density and electron temperature of plasma can be controlled by feeding in different frequencies from the second port. In this way, different frequencies can selectively drive the plasma characteristics. The OES (optical emission spectroscopy) results confirmed that the particles with different energy levels showed different responses at different frequencies. The comparison of the characteristics was carried out between the single-frequency microwave plasma and the dual-frequency microwave plasma in the same device. The presented device is interesting because it allows to some extent to regulate such plasma parameters as electron density and electron temperature. The range of these parameters can be extended by using microwave sources with adjustable power and frequency.

2.2. Cold Plasma Treatment of Melanoma Cells

Yun-Hsuan Chen et al. in [2] showed the results of application of cold atmospheric pressure plasma for treatment of melanoma (B16F10)-skin cancer cells (with and without catalase enzyme in vitro). The generated cold plasma was characterized electrically and

Citation: Jasiński, M. The Applications of Plasma Techniques II. *Appl. Sci.* **2022**, *12*, 3683. https://doi.org/10.3390/app12073683

Received: 29 March 2022
Accepted: 31 March 2022
Published: 6 April 2022

spectroscopically. Biological data showed that the plasma inactivated cancer cells but not non-malignant cells. The authors state that the presence of the CAT enzyme confirmed that the reactive oxygen and nitrogen species (RONS) play key roles in inactivating the melanoma cancer cells. The authors conclude that the cold atmospheric plasma is a promising tool to overcome certain cancerous and precancerous conditions in dermatology, and that detailed investigation is still needed to understand the mechanisms underlying the presented results. Humanity has been fighting cancer for a long time with a limited success, so the presented cancer-fighting technique seems to be important.

2.3. Vortex Breakdown Control by the Plasma Swirl Injector

Gang Li et al. in [3] proposed and tested the concept of using the plasma swirler to control vortex breakdown. The presented plasma swirler with a helical shape was adopted to control the vortex breakdown. The plasma swirler with the electrode placed in the streamwise direction was designed by the authors. The plasma actuation, in affecting the onset and development of the vortex breakdown was captured and analyzed using particle image velocimetry (PIV) technique. The flow field measurement demonstrated that the plasma actuation was effective in controlling the development of vortex. The authors conclude that the method being proposed here may represent an attractive way of controlling vortex breakdown using a small amount of energy input without a moving or intrusive part, and the plasma actuation offers great flexibility in flow and combustion control.

2.4. Influence of Ag Electrodes Asymmetry Arrangement on Their Erosion Wear and Nanoparticle Synthesis in Spark Discharge

Kirill Khabarov et al. in [4] investigated the effect of the asymmetry arrangement of Ag electrodes on them erosive wear and presented the synthesis of nanoparticles (NPs) during a spark discharge. The two types of discharge current pulses were studied: oscillation-damped, in which the electrodes changed their polarities during a single discharge, and unipolar, in which the electrodes had a given polarity during the discharge. The used electrodes in the form of rods, one of which had a gas supply hole, were installed coaxially. The authors demonstrated that it is possible to control the size and concentration of synthesized nanoparticles by changing the degree of the electrodes asymmetry by setting their end faces at a certain angle. With an increase in the degree of the electrodes asymmetry, larger nanoparticles (with sizes greater than 40 nm) appeared in the aerosol composition and their agglomeration increased. The results presented in this article can help spark discharge users to optimize the placement of discharge electrodes.

2.5. Impact of the Samples Surface State on the Glow Discharge Stability in the Metals Treatment and Welding Processes

Maksym Bolotov et al. in [5] presented the results of a study of the effect of sample surface condition on the stability of glow discharge in metalworking and welding processes. The main objective of this study was to investigate the effect of cathode macro and micro relief on the existence of stable glow discharge in metalworking and diffusion welding processes. It was determined analytically and supported experimentally that the stability of the glow discharge is mainly affected by the generated sharp protrusions on the cathode surface due to the pretreatment of the samples by machining before welding. It was found that the increasing the cathode surface roughness from 10–15 μm to 60–80 μm led to the rapid decreasing the region of the limiting pressure of the stable glow discharge from 1.33–13.3 kPa to 1.33–5.3 kPa. The results presented in this article may be helpful for glow discharge researchers to optimize the working gas pressure and surface roughness of discharge electrodes.

2.6. Surface Discharge Mechanism on Epoxy Resin in Electronegative Gases and Its Application

Herie Park et al. in [6] presented a work on the mechanism of surface discharge on epoxy resin in electronegative gases and its application. The authors showed the characteristics of surface discharges in compressed insulating gases, including sulfur hexafluoride (SF_6), dry air and N_2 under a non-uniform electric field. The experiments were conducted with the gas pressure from 0.1 to 0.6 MPa using samples of epoxy dielectrics under an AC voltage. The experimental results showed that the surface insulation performance improved significantly using insulating gases containing electronegative gases, such as SF_6 and dry air. Among the various gases, SF_6 and dry air, which are electronegative gases, showed better insulating properties compared to N_2 due to their electron attachment capacity. The influence of electronegative gases on the surface ignition voltages, which vary with the pressure in these gases, has been analyzed in detail through the processes of electron attachment and detachment. The authors conclude that the physical information obtained from the results of this study can be used to provide improved surface insulation performance by using an insulating gas mixing technique in the design of SF_6-free HV equipment surface insulation.

3. Conclusions

The collection of articles discussed above covers various types of discharges and various processes. The discharges presented include, for example, microwave, spark, glow, or surface discharges. The characterizations of the sources of these discharges, the parameters of the generated plasmas, as well as the applications of these plasmas are discussed. The applications include, for example, the synthesis of nanoparticles or the treatment of skin cancer cells. I hope that the presented articles will be valuable for readers representing the world of science, medicine, and technology.

Acknowledgments: Thank you to all the authors and reviewers for their valuable contributions to this Special Issue (The Applications of Plasma Techniques II) of Applied Sciences. Congratulations to the management and all MDPI staff for the editorial support that contributed to the success of this project. I would especially like to thank Seraina Shi (Menaging Editor from MDPI Branch Office, Tianjin) for direct cooperation and assistance.

Conflicts of Interest: The authors declare no conflict of interest.

References

1. Chen, C.; Fu, W.; Zhang, C.; Lu, D.; Han, M.; Yan, Y. Dual-frequency microwave plasma source based on microwave coaxial transmission line. *Appl. Sci.* **2021**, *11*, 9873. [CrossRef]
2. Chen, Y.-H.; Hsieh, J.-H.; Wang, I.-T.; Jheng, P.-R.; Yeh, Y.-Y.; Lee, J.-W.; Bolouki, N.; Chuang, E.-Y. Transferred cold atmospheric plasma treatment on melanoma skin cancer cells with/without catalase enzyme In Vitro. *Appl. Sci.* **2021**, *11*, 6181. [CrossRef]
3. Li, G.; Jiang, X.; Du, W.; Yang, J.; Liu, C.; Mu, Y.; Xu, G. Vortex breakdown control by the plasma swirl injector. *Appl. Sci.* **2021**, *11*, 5537. [CrossRef]
4. Khabarov, K.; Urazov, M.; Lizunova, A.; Kameneva, E.; Efimov, A.; Ivanov, V. Influence of Ag electrodes asymmetry arrangement on their erosion wear and nanoparticle synthesis in spark discharge. *Appl. Sci.* **2021**, *11*, 4147. [CrossRef]
5. Bolotov, M.; Bolotov, G.; Stepenko, S.; Ihnatenko, P. Impact of the samples' surface state on the glow discharge stability in the metals' treatment and welding processes. *Appl. Sci.* **2021**, *11*, 1765. [CrossRef]
6. Park, H.; Lim, D.-Y.; Bae, S. Surface discharge mechanism on epoxy resin in electronegative gases and its application. *Appl. Sci.* **2020**, *10*, 6673. [CrossRef]

 *applied sciences*

Technical Note

Dual-Frequency Microwave Plasma Source Based on Microwave Coaxial Transmission Line

Chi Chen, Wenjie Fu *, Chaoyang Zhang, Dun Lu, Meng Han and Yang Yan

School of Electronic Science and Engineering, University of Electronic Science and Technology of China, Chengdu 610054, China; chenchi@std.uestc.edu.cn (C.C.); chaoyangzhang@std.uestc.edu.cn (C.Z.); ludun@std.uestc.edu.cn (D.L.); 201711040132@std.uestc.edu.cn (M.H.); yanyang@uestc.edu.cn (Y.Y.)

* Correspondence: fuwenjie@uestc.edu.cn

Featured Application: In various microwave plasma fields, such as Microwave Plasma Chemical Vapor Deposition (MPCVD), the dual-frequency plasma source could be used to control plasma characteristics flexibly.

Abstract: A dual-frequency plasma source has many advantages in applications. In this paper, a dual-frequency microwave plasma source is presented. This microwave plasma source is based on a coaxial transmission line without the resonator, and it can be operated in a wide band frequency region. Two microwaves are inputted from two ports into the plasma reactor: one is used firstly to excite the plasma and the other one is used to adjust plasma characteristics. Based on the COMSOL Multiphysics simulation, the experiment is carried out. In the experimental investigation, the plasma electron density and electron temperature can be controlled, respectively, by feeding in different frequencies from the second port, causing the particles at different energy levels to present different frequencies. This exploratory research improves the operation frequency of dual-frequency microwave plasma sources from RF to microwave.

Keywords: microwave plasma; dual-frequency plasma; electron temperature; electron density

Citation: Chen, C.; Fu, W.; Zhang, C.; Lu, D.; Han, M.; Yan, Y. Dual-Frequency Microwave Plasma Source Based on Microwave Coaxial Transmission Line. *Appl. Sci.* **2021**, *11*, 9873. https://doi.org/10.3390/app11219873

Academic Editor: Mariusz Jasiński

Received: 6 October 2021
Accepted: 19 October 2021
Published: 22 October 2021

1. Introduction

To control plasma characteristics flexibly, the dual-frequency plasma source has been proposed and investigated [1–8]. The dual-frequency plasma source is a hybrid source, in which one frequency is chosen to be much higher than the other in order to achieve independent control of ion bombardment and electron density [9]. Research shows that, in plasma etching, dual-frequency operating could reduce particle contamination in the plasma reactor [10], and in the plasma-enhanced chemical vapor deposition (PECVD), dual-frequency operating could improve the film stress, step coverage, chemical composition, and film stability [11–14].

In previous studies, the exciting frequency for dual-frequency plasma sources mostly were mostly radio frequency (RF), such as 13.56 MHz, 27.12 MHz, 320 MHz, 340 kHz, and 40 kHz, and the structures of the plasma reactor are mostly capacitively coupled plasma (CCP) and/or inductively coupled plasma (ICP). In recent years, more and more microwave plasma sources have been proposed and applied [15–18]. However, dual-frequency microwave plasma sources are rarely investigated. For most microwave plasma sources, the microwave is propagated through a waveguide, and the plasma is excited in a resonator. Different from CCP and ICP reactor structures [19], the size of the microwave resonator is dependent on microwave frequency, and it is difficult to resonate two non-harmonic frequencies in one resonator.

Here, a microwave coaxial transmission line is introduced as the reactor for a dual-frequency microwave plasma source. In the coaxial transmission line, the electromagnetic wave mode of the microwave is TEM mode, and ultra-wide band electromagnetic waves, from low-frequency wave to millimeter wave, could be inputted and propagated. The 2450 and 915 MHz microwaves, which are widely used in industry, are adopted in this investigation. There are two ports to input the microwaves. The 2450 MHz microwave is inputted from one port to excite plasma, and the 915 MHz microwave is inputted from another to adjust the plasma characteristics. To investigate the characteristics of dual-frequency microwave plasma, the microwave is inputted from two ports to excite the plasma, and a comparison is conducted between single-frequency microwave plasma and dual-frequency microwave plasma in the same reactor.

2. Experiment Design

The schematic diagram for the presented dual-frequency microwave plasma source based on the microwave coaxial transmission line is shown in Figure 1.

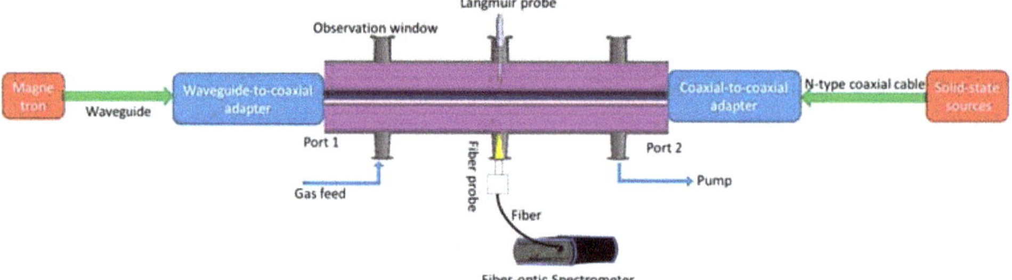

Figure 1. Schematic of the coaxial line plasma driven by the dual microwave sources.

The plasma reactor is a coaxial line, whose outer diameter is 150 mm and the inner diameter is 8 mm. A quartz glass tube surrounds the inner conductor. Between the outer conductor and the quartz glass tube, there is a vacuum region in which the plasma is excited. There are two microwave ports at the two ends of the coaxial transmission line. One is used to input the 2450 MHz microwave, and another is used to input the 915 MHz microwave. A magnetron microwave power generator (ASTeX AX2110) is used to generate the continuous-wave 2450 MHz microwave. An isolation circulator with a water load in an isolated way is connected to the magnetron microwave power generator. In this way, the microwave feedback will be entirely absorbed by the water load. An ASTeX SmartMatch is used to match the impedance so that the maximum amount of power can be coupled with the plasma. The 2450 MHz microwave is transmitted from the generator to the plasma reactor by the waveguide. A waveguide-to-coaxial adapter is designed and installed to guide the microwave from the waveguide into the coaxial transmission line plasma reactor through Port 1 [15]. A custom-made solid-state microwave power source (Wattsine Electronic Technology Co., Ltd. Chengdu China) is used to generate a continuous-wave 915 MHz microwave. The 915 MHz microwave is transmitted by N-type coaxial cables. The incident and reflected powers are monitored by a bi-directional coaxial coupler and Keysight power meter (Narda N1914A). A coaxial-to-coaxial adapter is designed and installed to guide the microwave from the N-type coaxial cable into the coaxial transmission line plasma reactor through Port 2. A coaxial low-pass filter is installed into N-type coaxial cables to avoid the solid-state microwave power source affected by 2450 MHz.

There are several KF flanges on the outer conductor, which are used to connect a vacuum pump, gas feed mass flow meter, and diagnostic devices. A Langmuir probe and an optical emission spectrometry (OES) are used to diagnose and investigate the plasma characteristics from two KF flanges in the center of the plasma reactor.

The experimental processes are as follows: First, pump the pressure in the reactor below 10 Pa by a vacuum pump; second, adjust the argon flow rate at 15~20 sccm to stabilize the discharge pressure at 80 Pa by the mass flow meter; then, turn on the microwave source at Port 1 to excite plasma discharge; finally, turn on the microwave source at Port 2 to adjust plasma characteristics and measure the plasma parameters.

As shown in Figure 1, the coaxial transmission line is a non-resonant structure. Thus, ultra-wide band electromagnetic waves could be inputted into the reactor. However, it would only be discharged at the area near the input Port 1, and not at the whole reactor. To produce a dual-frequency operation, the plasma must be affected by microwaves from both ports. Therefore, the plasma excited by 2450 MHz from Port 1 should be distributed in the whole reactor. To obtain the plasma distribution in the reactor, COMSOL Multiphysics is used to simulate the plasma distribution at different microwave powers from Port 1.

In this investigation, the argon plasma considered in the simulated model contains only electrons (e), positively charged argon ions (Ar^+), metastable-state argon (Ars), and argon (Ar) atoms. For these species, the main transportation equations are provided in Table 1 [20].

Table 1. Important collision processes in the argon discharge.

No.	Reaction	Type	Energy Loss $\Delta\varepsilon$ (eV)
1	$e + Ar = e + Ar$	Elastic	
2	$e + Ar = e + Ars$	Excitation	11.5
3	$e + Ars = e + Ar$	Superelastic	−11.5
4	$e + Ars = 2e + Ar^+$	Ionization	15.8
5	$e + Ar = 2e + Ar^+$	Ionization	4.24
6	$Ars + Ars = e + Ar + Ar^+$	Penning ionization	
7	$Ars + Ar = Ar + Ar$	Metastable quenching	

Stepwise ionization (Reaction 5 in Table 1) plays an important role in sustaining low-pressure argon discharges. Excited argon atoms are consumed via superelastic collisions with electrons, quenching with neutral argon atoms, ionization, or Penning ionization where two metastable argon atoms react to form a neutral argon atom, an argon ion, and an electron. Reaction 7 is responsible for the heating of the gas. The 11.5 eV of energy, which was consumed in creating the electronically excited argon atom, returns to the gas as thermal energy when the excited metastable-state argon atoms quenching.

In the simulation, the electron energy distribution function (EEDF) is approximately set as the Maxwellian distribution function, due to the fact that the discharge pressure 80 Pa is greater than 6.66 Pa [21,22]. Although publications show that the plasma densities in simulations as Maxwellian distribution are higher than experimental measurements, they also show that the plasma distributions in simulations present good agreement with experimental measurements [23–26]. Therefore, the Maxwellian distribution function is still useful, and the plasma distributions at different exciting powers are normalized, presented in Figure 2.

As shown in Figure 2, the plasma discharge area is spread with increasing Port 1 microwave power. When the inputted 2450 MHz microwave power is at 500 W, the electron density near Port 2 is much lower than that near Port 1. When the inputted 2450 MHz microwave power is at 600 W, the electron density near Port 2 is a little lower than that near Port 1, but the plasma distribution is still non-uniform. When the inputted 2450 MHz microwave power is at 700 W, the electron density near Port 1 is nearly the same as that near Port 2, and in the area between Port 1 and Port 2, the plasma distribution is approximately uniform. While electron density is sufficiently affected by microwave power, the electron temperature is almost unaffected by microwave power changes. Therefore, the microwave power at Port 1 was set at 700 W to excite the plasma.

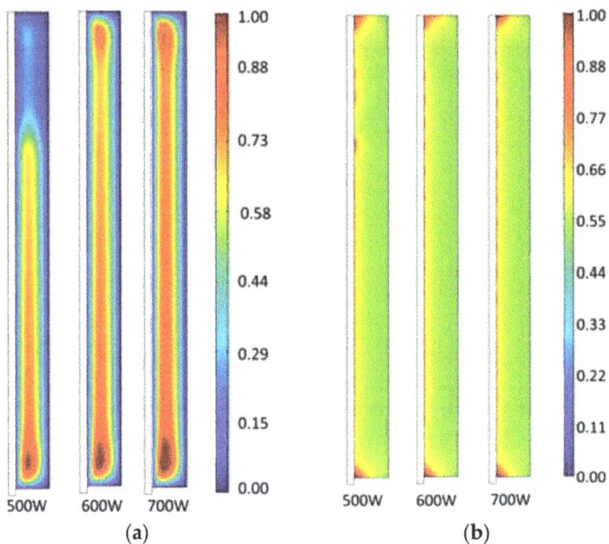

Figure 2. COMSOL simulation of the electron density and temperature distributions at a fixed pressure of 80 Pa under Port 1 2450 MHz microwave power: (**a**) electron density (**b**) electron temperature.

3. Experiment Results

Based on the simulation results, the experiment was carried out in accordance with processes listed in Section 2. Firstly, the 700 W 2450 MHz microwave was inputted into the reactor from Port 1 at a pressure of 80 Pa. The plasma can be excited with the assistance of an automatic impedance matchbox. In the center of the plasma reactor, the electron density and temperature measured by the Langmuir probe were $n_e = 1.56 \times 10^{16}\ \text{m}^{-3}$ and $Te = 2.46\ (\text{eV})$.

For microwave-excited plasma, electrons oscillate among the heavy immobile ions at the plasma electron frequency. At this time, if another microwave with the frequency below the plasma electron frequency is inputted from another port, the inertia of the electrons is sufficiently low for them to respond to the electric field in the incident electromagnetic wave, and the electron can, therefore, absorb energy from it. If the frequency of the incident microwave is higher than the electron plasma frequency, the inertia of an electron would be too high to enable it to respond fully to the incident microwave. Therefore, the interaction of the microwave with individual electrons is relatively insignificant. Thus, different interaction modes of microwaves with electrons can be achieved by controlling the frequencies of the second incident microwaves [27].

The electron plasma frequency is a function of the electron number density and is given by [27].

$$\omega_{\text{pe}} = 2\pi v_{pe} = \sqrt{\frac{n_e e^2}{\varepsilon_0 m_e}}\ \ (\text{rad/s}) \quad (1)$$

where v_{pe} is the electron number density in Hz. The n_e is the electron density, the ε_0 is the permittivity of free space, and the m_e is the mass of an electron. According to Equation (1), for $n_e = 1.56 \times 10^{16}\ \text{m}^{-3}$, the electron plasma frequency is $\nu_e = 1123$ MH, which is located between 2450 MHz and 915 MHz. Then, if the input microwave from Port 2 is 2450 or 915 MHz, the plasma would show different responding behaviors.

To compare with the influence of the 915 MHz microwave, a 2450 MHz solid-state source was used instead of the 915 MHz solid-state source and the low-pass filter for comparative experiments. In the experiment, 915 and 2450 MHz microwaves were inputted from Port 2 to investigate the plasma characteristics changes, respectively. Figure 3 shows the effect of microwave inputted from Port 2 on the electron density and electron temperature measured by the Langmuir probe.

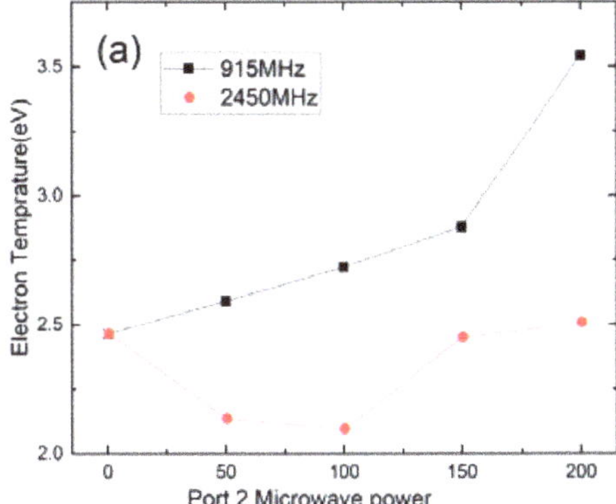

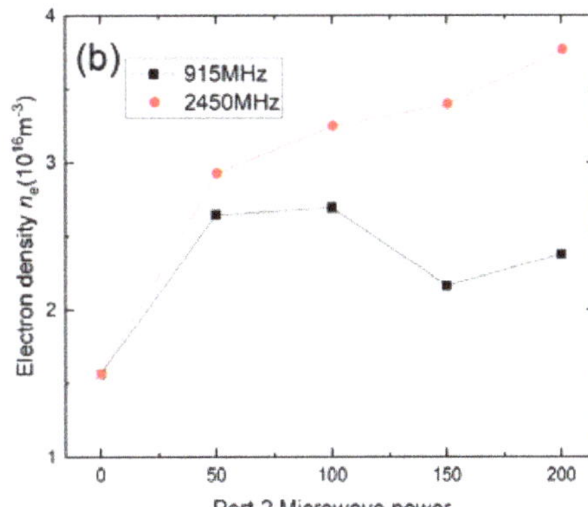

Figure 3. Effect of the microwave inputted from port 2 measured by the Langmuir probe: (**a**) electron temperature; (**b**) electron density.

Figure 3a shows that, when the second microwave was inputted from Port 2, the plasma electron temperature gradually increased as the 915 MHz microwave power increased and did not change sufficiently as the 2450 MHz microwave power increased. In contrast, Figure 3b shows that, when the second microwave was inputted from Port 2, the plasma electron density did not change sufficiently as the 915 MHz microwave power increased, but it gradually increased as the 2450 MHz microwave power increased.

This means that the second incident microwaves with frequencies below the plasma electron frequency could increase the electron temperature and affect the electron density indistinctly, and the second incident microwaves with frequencies above the plasma electron frequency could increase the electron density and affect the electron temperature indistinctly.

To investigate the physical mechanism of these phenomena, the OES results at different conditions were analyzed, and the OES data were all obtained at 10-ms integral times. The spectrum at only 700 W 2450 MHz microwave that was inputted from Port 1 acquired by the OES method is presented in Figure 4.

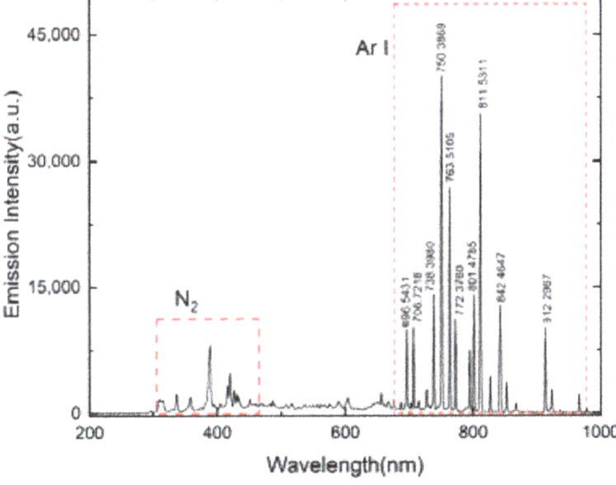

Figure 4. Spectrum of the coaxial line plasma driven by the 2450 MHz microwave from Port 1 at a power of 700 W and pressure of 80 Pa.

The spectrum result presents two main particles in plasma: nitrogen molecule (N_2) and Ar. The N_2 comes from air residue. In the experiment, the intensity of the Ar I line group was much higher than the others. In the Ar I line group, the characteristic peaks at Ar I 750.3869 nm, and Ar I 811.5311 nm were quite obvious with the highest intensity.

The OES data at the 200 W–915 MHz microwave that was inputted from Port 2 and at the 200 W–2450 MHz microwave that was inputted from Port 2 are presented in Figure 5a,b, respectively.

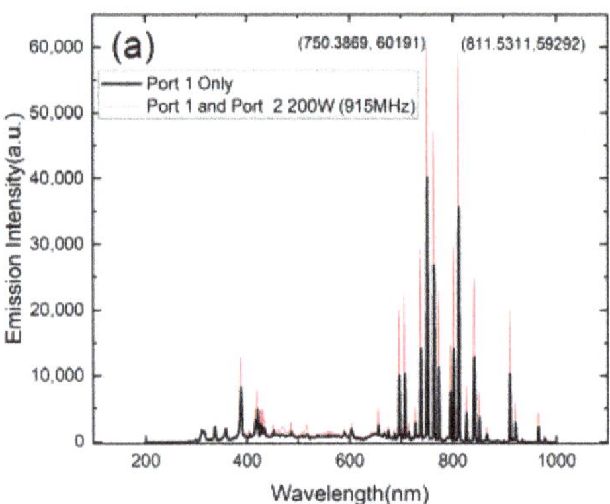

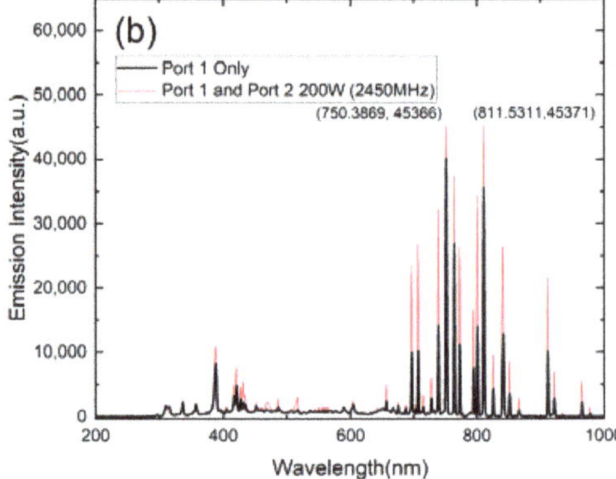

Figure 5. Spectrum of microwave inputted from Port 2 at the power of 200 W: (**a**) 915 MHz microwave; (**b**) 2450 MHz microwave.

In Figure 5, it is obviously seen that the intensity of the Ar I characteristic peaks was sufficiently improved when both 915 and 2450 MHz microwaves were inputted, and different Ar I characteristic peaks show different changes. The intensity of the two highest Ar I characteristic peaks (750.3869 and 811.5311 nm) at the 915 MHz microwave that was inputted from Port 2 were higher than that of the 2450 MHz microwave that was inputted from Port 2. Meanwhile, the intensity of other Ar I characteristic peaks (such as 696.5431 nm, 706.7218 nm, 772.3760 nm) at the 915 MHz microwave that was inputted from Port 2 are lower than that of the 2450 MHz microwave that was inputted from Port 2.

The intensity changes of several Ar I characteristic peaks versus microwave power that were inputted from Port 2 are shown in Figure 6.

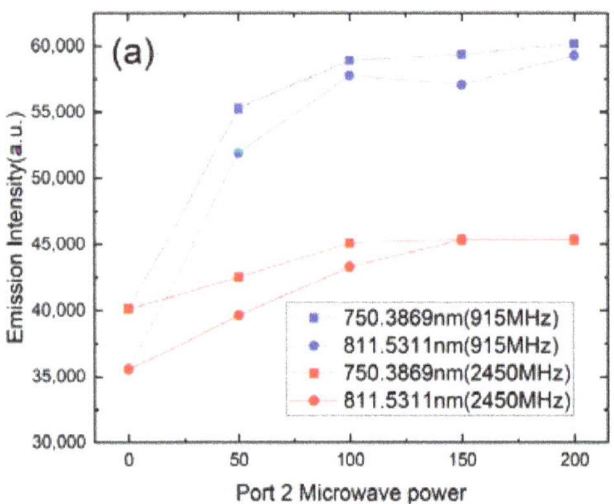

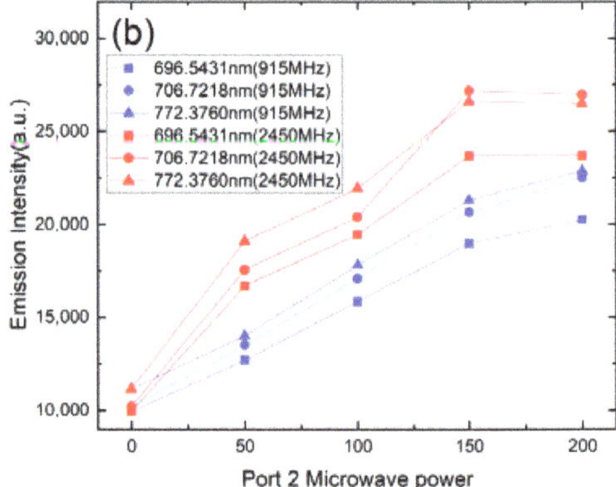

Figure 6. Intensity changes of Ar I characteristic peaks versus microwave power inputted from port 2 (**a**) 750.3869 and 811.5311 nm; (**b**) 696.5431, 706.7218 and 772.3760 nm.

The OES results show that when the intensity of the characteristic peaks is higher at only 700 W 2450 MHz that was inputted from Port 1, the intensity increase is higher at 915 MHz that was inputted from Port 2 than at 2450 MHz that was inputted from Port 2 (Table S1, Figure S1). This indicates that particles at different energy levels presented different responses to different frequencies. In addition, the particles at different energy levels contributed different ratios in plasma electron density and electron temperature, thus, different frequencies could selectively control the plasma characteristics.

4. Summary and Discussion

Based on the COMSOL Multiphysics simulation, a dual-frequency microwave plasma source is proposed and investigated. This dual-frequency microwave plasma source is based on the coaxial transmission line, and 2450 and 915 MHz microwaves are utilized in this study.

In the experiment, the plasma was excited by a 2450 MHz microwave from Port 1 first, and then a 915 MHz microwave was inputted from Port 2 to adjust the plasma characteristics. The measured experiment results by the Langmuir probe show that the electron temperature gradually increases as the 915 MHz microwave power increases, and there is little change in electron density. In contrast, if there is a 2450 MHz microwave that is inputted from Port 2, the electron density increases as the microwave power increases, and there is little change in the electron temperature. These phenomena of dual-frequency microwave plasma are similar to dual-frequency RF plasma.

The OES method is used to investigate the physical mechanism of these phenomena, and the OES results show that the particles at different energy levels presented different responses to different frequencies. Thus, different frequencies could selectively control the plasma characteristics. If a high-power wide bandwidth frequency tunable microwave source is used as the power source from Port 2, the plasma electron density and electron temperature would be controlled, respectively and precisely, in the dual-frequency microwave plasma source. This exploratory research would improve the operation frequency of dual-frequency microwave plasma sources from RF to microwave.

Supplementary Materials: The following are available online at https://www.mdpi.com/article/10.3390/app11219873/s1, Table S1: Intensity of Ar I characteristic peaks, Figure S1: Intensity of Ar I characteristic peaks versus microwave power inputted from port 2 under pressure of 80 Pa and 700 W 2450 MHz microwave at Port 1.

Author Contributions: C.C. and W.F. contributed to the overall study design, analysis, and writing of the manuscript. C.Z., D.L., M.H. and Y.Y. provided technical support and revised the manuscript. All authors have read and agreed to the published version of the manuscript.

Funding: This work was supported by the National Key Research and Development Program of China under 2019YFA0210202, the National Natural Science Foundation of China under Grant 61971097 and 62111530054.

Conflicts of Interest: The authors declare no conflict of interest.

References

1. Shin, K.S.; Sahu, B.B.; Han, J.G.; Hori, M. Utility of dual frequency hybrid source for plasma and radical generation in plasma enhanced chemical vapor deposition process. *Jpn. J. Appl. Phys.* **2015**, *54*, 76201. [CrossRef]
2. van de Ven, E.P.; Connick, I.W.; Harrus, A.S. Advantages of dual frequency PECVD for deposition of ILD and passivation films. In Proceedings of the Seventh International IEEE Conference on VLSI Multilevel Interconnection, Santa Clara, CA, USA, 12–13 June 1990.
3. Lafleur, T.; Delattre, P.A.; Johnson, E.V.; Booth, J.P. Separate control of the ion flux and ion energy in capacitively coupled radio-frequency discharges using voltage waveform tailoring. *Appl. Phys. Lett.* **2012**, *101*, 124104. [CrossRef]
4. Curley, G.A.; Marić, D.; Booth, J.; Corr, C.S.; Chabert, P.; Guillon, J. Negative ions in single and dual frequency capacitively coupled fluorocarbon plasmas. *Plasma Sources Sci. Technol.* **2007**, *16*, S87–S93. [CrossRef]
5. Lee, J.K.; Manuilenko, O.V.; Babaeva, N.Y.; Kim, H.C.; Shon, J.W. Ion energy distribution control in single and dual frequency capacitive plasma sources. *Plasma Sources Sci. Technol.* **2005**, *14*, 89–97. [CrossRef]

6. Booth, J.P.; Curley, G.; Marić, D.; Chabert, P. Dual-frequency capacitive radiofrequency discharges: Effect of low-frequency power on electron density and ion flux. *Plasma Sources Sci. Technol.* **2010**, *19*, 15005. [CrossRef]
7. Schulze, J.; Schüngel, E.; Czarnetzki, U.; Donkó, Z. Optimization of the electrical asymmetry effect in dual-frequency capacitively coupled radio frequency discharges: Experiment, simulation, and model. *J. Appl. Phys.* **2009**, *106*, 63307. [CrossRef]
8. Maeshige, K.; Washio, G.; Yagisawa, T.; Makabe, T. Functional design of a pulsed two-frequency capacitively coupled plasma in CF[sub 4]/Ar for SiO[sub 2] etching. *J. Appl. Phys.* **2002**, *91*, 9494. [CrossRef]
9. Piallat, F.; Vallée, C.; Gassilloud, R.; Michallon, P.; Pelissier, B.; Caubet, P. PECVD RF versus dual frequency: An investigation of plasma influence on metal–organic precursors' decomposition and material characteristics. *J. Phys. D Appl. Phys.* **2014**, *47*, 185201. [CrossRef]
10. Kim, M.; Cheong, H.; Whang, K. Particle formation and its control in dual frequency plasma etching reactors. *J. Vac. Sci. Technol. A Vac. Surf. Film.* **2015**, *33*, 41303. [CrossRef]
11. Cianci, E.; Schina, A.; Minotti, A.; Quaresima, S.; Foglietti, V. Dual frequency PECVD silicon nitride for fabrication of CMUTs' membranes. *Sens. Actuators A Phys.* **2006**, *127*, 80–87. [CrossRef]
12. Pearce, C.W.; Fetcho, R.F.; Gross, M.D.; Koefer, R.F.; Pudliner, R.A. Characteristics of silicon nitride deposited by plasma-enhanced chemical vapor deposition using a dual frequency radio-frequency source. *J. Appl. Phys.* **1992**, *71*, 1838. [CrossRef]
13. Tarraf, A.; Daleiden, J.; Irmer, S.; Prasai, D.; Hillmer, H. Stress investigation of PECVD dielectric layers for advanced optical MEMS. *J. Micromechanics Microeng.* **2003**, *14*, 317–323. [CrossRef]
14. Zhou, X.; Tan, X.; Lv, Y.; Wang, Y.; Li, J.; Liang, S.; Zhang, Z.; Feng, Z.; Cai, S. 128-pixel arrays of 4H-SiC UV APD with dual-frequency PECVD SiNx passivation. *Opt. Express* **2020**, *28*, 29245. [CrossRef] [PubMed]
15. Chen, C.; Fu, W.; Zhang, C.; Lu, D.; Han, M.; Yan, Y. Langmuir Probe Diagnostics with Optical Emission Spectrometry (OES) for Coaxial Line Microwave Plasma. *Appl. Sci.* **2020**, *10*, 8117. [CrossRef]
16. Cha, J.; Kim, S.; Lee, H. A Linear Microwave Plasma Source Using a Circular Waveguide Filled with a Relatively High-Permittivity Dielectric: Comparison with a Conventional Quasi-Coaxial Line Waveguide. *Appl. Sci.* **2021**, *11*, 5358. [CrossRef]
17. Salgado-Meza, M.; Martínez-Rodríguez, G.; Tirado-Cantú, P.; Montijo-Valenzuela, E.E.; García-Gutiérrez, R. Synthesis and Properties of Electrically Conductive/Nitrogen Grain Boundaries Incorporated Ultrananocrystalline Diamond (N-UNCD) Thin Films Grown by Microwave Plasma Chemical Vapor Deposition (MPCVD). *Appl. Sci.* **2021**, *11*, 8443. [CrossRef]
18. Wiktor, A.; Hrycak, B.; Jasiński, M.; Rybak, K.; Kieliszek, M.; Kraśniewska, K.; Witrowa-Rajchert, D. Impact of Atmospheric Pressure Microwave Plasma Treatment on Quality of Selected Spices. *Appl. Sci.* **2020**, *10*, 6815. [CrossRef]
19. Zhang, Y.; Kushner, M.J.; Sriraman, S.; Marakhtanov, A.; Holland, J.; Paterson, A. Control of ion energy and angular distributions in dual-frequency capacitively coupled plasmas through power ratios and phase: Consequences on etch profiles. *J. Vac. Sci. Technol. A* **2015**, *33*, 31302. [CrossRef]
20. Kim, D.; Jeong, Y.; Shon, Y.; Kwon, D.; Jeon, J.; Choe, H. Modified Fluid Simulation of an Inductively Coupled Plasma Discharge. *Appl. Sci. Converg. Technol.* **2019**, *28*, 221–225. [CrossRef]
21. Franz, G. *Low Pressure Plasmas and Microstructuring Technology*; Springer: Berlin/Heidelberg, Germany, 2009; pp. 188–190.
22. Rebiai, S.; Bahouh, H.; Sahli, S. 2-D simulation of dual frequency capacitively coupled helium plasma, using COMSOL multiphysics. *IEEE Trans. Dielectr. Electr. Insul.* **2013**, *20*, 1616–1624. [CrossRef]
23. Ochoa Brezmes, A.; Breitkopf, C. Simulation of inductively coupled plasma with applied bias voltage using COMSOL. *Vacuum* **2014**, *109*, 52–60. [CrossRef]
24. Lei, F.; Li, X.; Liu, Y.; Liu, D.; Yang, M.; Yu, Y. Simulation of a large size inductively coupled plasma generator and comparison with experimental data. *AIP Adv.* **2018**, *8*, 15003.
25. Lymberopoulos, D.P.; Economou, D.J. Modeling and simulation of glow discharge plasma reactors. *J. Vac. Sci. Technol. A Vac. Surf. Film.* **1994**, *12*, 1229–1236. [CrossRef]
26. Shahbazian, A.; Salem, M.K.; Ghoranneviss, M. Simulation by COMSOL of Effects of Probe on Inductively Coupled Argon Plasma. *Braz. J. Phys.* **2021**, *51*, 351. [CrossRef]
27. Reece Roth, J. Principles. In *Industrial Plasma Engineering*; CRC Press: Boca Raton, FL, USA, 1995; Volume 1.

Article

Transferred Cold Atmospheric Plasma Treatment on Melanoma Skin Cancer Cells with/without Catalase Enzyme In Vitro

Yun-Hsuan Chen [1,†], Jang-Hsing Hsieh [2,3,†], I-Te Wang [4,†], Pei-Ru Jheng [1], Yi-Yen Yeh [1], Jyh-Wei Lee [2,3], Nima Bolouki [3,*] and Er-Yuan Chuang [1,5,*]

1 Graduate Institute of Biomedical Materials and Tissue Engineering, International Ph.D. Program in Biomedical Engineering, College of Biomedical Engineering, Taipei Medical University, Taipei 110, Taiwan; ttpss90128@gmail.com (Y.-H.C.); m120104012@tmu.edu.tw (P.-R.J.); pineapple2821@yahoo.com.tw (Y.-Y.Y.)
2 Department of Materials Engineering, Ming Chi University of Technology, New Taipei City 243, Taiwan; jhhsieh@mail.mcut.edu.tw (J.-H.H.); jefflee@mail.mcut.edu.tw (J.-W.L.)
3 Center for Plasma and Thin Film Technologies, Ming Chi University of Technology, New Taipei City 243, Taiwan
4 Department of Obstetrics and Gynecology, Taipei Medical University Hospital, Taipei 110, Taiwan; b8301120@yahoo.com.tw
5 Cell Physiology and Molecular Image Research Center, Taipei Medical University, Wan Fang Hospital, Taipei 116, Taiwan
* Correspondence: bolouki@mail.mcut.edu.tw (N.B.); eychuang@tmu.edu.tw (E.-Y.C.)
† These authors contributed equally to this work.

Citation: Chen, Y.-H.; Hsieh, J.-H.; Wang, I-T.; Jheng, P.-R.; Yeh, Y.-Y.; Lee, J.-W.; Bolouki, N.; Chuang, E.-Y. Transferred Cold Atmospheric Plasma Treatment on Melanoma Skin Cancer Cells with/without Catalase Enzyme In Vitro. *Appl. Sci.* **2021**, *11*, 6181. https://doi.org/10.3390/app11136181

Academic Editor: Mariusz Jasiński

Received: 13 May 2021
Accepted: 30 June 2021
Published: 2 July 2021

Publisher's Note: MDPI stays neutral with regard to jurisdictional claims in published maps and institutional affiliations.

Abstract: Cold atmospheric plasma (CAP) is a promising tool to overcome certain cancerous and precancerous conditions in dermatology. A scheme of transferred CAP was first developed to treat melanoma (B16F10) skin cancer cells as well as non-malignant (L929) cells in vitro. CAP was transferred using a silicone tube with a jet system that was developed and was assessed as to whether it could generate reactive oxygen and nitrogen species (RONS) at near-room temperature. The transferred CAP was characterized electrically and spectroscopically. Biological data showed that the transferred CAP killed cancer cells but not non-malignant (L929) cells. Plasma treatment was effective with a time duration of 30 s, whereas non-malignant (L929) cells were less damaged during plasma treatment. In addition, catalase (CAT) enzyme was applied to neutralize and detoxify the RONS generated by the transferred CAP. These findings suggest that transferred CAP can be considered a melanoma cancer therapy.

Keywords: melanoma cell (B16F10); plasma cancer therapy; cold atmospheric plasma (CAP); transferred cold atmospheric plasma; reactive oxygen species (ROS); reactive nitrogen species (RNS); catalase

1. Introduction

Cancer is one of the major diseases and a leading cause of death worldwide. Among the various forms of cancer, melanoma skin cancer is regarded as one of the deadliest cancers. A survey revealed that melanoma skin cancer is rising faster than other types of cancers particularly in developed countries [1], which is probably due to the increasing ultraviolet (UV) radiation and holes in the ozone layer [2].

Advancements in cancer therapies are specifically due to innovative technologies from various fields of science and engineering which have improved diagnostic and treatment systems of patients. In the case of skin cancer cells, several treatment methods are suggested; however, the suggested treatments include undesired side effects. For instance, surgery as a primary treatment is able to remove skin cancer cells, including some healthy cells around cancer cells. Such treatment may leave scarring, and in some cases may even be painful. In the case of immunotherapy, several types of chemo-drugs are used for treating skin cancer, but they have various unwanted side effects as well. Chemotherapy is known as a common option for treating skin cancer cells [3]. Since this treatment is drug-based, in addition to side effects such as hair loss and so on, it is regarded as an ineffective treatment due to drug resistance [4]. As reported in a previously published article, various skin cancer cell treatments were well categorized, including the side effects of conventional treatments for skin cancer cells [5].

Radiation treatment uses high-energy radiation, such as x-rays or particles, to kill skin cancer cells. However, the side effects of radiation treatment are not usually limited to areas that receive radiation. Typical side effects may include changes in skin color, nausea, hair loss, and fatigue [6]. Suggested treatments for skin melanomas showed several advantages and disadvantages for clinical patients. Furthermore, all melanoma patients must undergo a screening process and get motivation or advice from dermatologists. They might be able to avoid experiencing possible side effects from the therapy. Determining how to prevent side effects during melanoma treatments and make melanoma patients comfortable is important and urgently needed. One of the novel melanoma treatments relies on plasma at atmospheric pressure.

In recent decades, plasma-based treatments were introduced in the field of medicine. Plasma is the fourth state of matter, including charged particles (ions and electrons), photons, and neutral atoms; it has a net neutral charge [7]. Plasma medicine has emerged as an interdisciplinary research field combining plasma physics, plasma chemistry, biology, and clinical medicine [8]. Plasma sources, in this field, are mainly focused on plasmas at atmospheric pressure that generate charged particles, radicals, excited species, and reactive oxygen (ROS) and nitrogen species (RNS). When interacting with ambient air, atmospheric-pressure plasma transfers the energy of particles of a noble gas with oxygen and nitrogen in ambient air; consequently, numerous ROS and RNS are generated such as oxygen radicals, nitrogen species, and so on. Here, we also merge ROS and RNS to call them RONS. The cytotoxicity effects of these species contribute to biomedical treatments [9–12], specifically in cancer therapies [13–15].

Generally, cold atmospheric plasma (CAP) sources are generated with various electrode configurations [16]. Herein, a dielectric barrier discharge (DBD)-based plasma jet configuration was employed, as it is more adaptable and safer for biomedical treatment. A specific configuration of the DBD-based plasma jet, which is called an extendable plasma jet, was utilized for melanoma cell treatment. Similar configurations were reported for endoscopic plasma applications [17], inner surface modification [18,19], and bio-targeting of human cell lines, such as A431 (skin carcinoma), HEK 293 (kidney embryonic cells), and A549 cell lines (human lung adenocarcinoma cells) [20]. In the case of biomedical treatment, the use of transferred CAPs, including helium and neon gases, were reported [17,20]; however, using a transferred CAP including argon gas, has not yet been mentioned for melanoma cancer cells. It should be noted that argon gas was just used for conventional plasma jet systems [20]. It has been found that the temperature of argon plasma with the configuration of the conventional plasma jet rose to around 40 °C. This heating problem is due to using argon gas [21] that would not be biocompatible. Whereas, using a scheme of transferred plasma, the gas temperature cools when the plasma gas is flowing from upstream to downstream in the jet system. In this case, the gas temperature is kept at a biocompatible level approaching room temperature to avoid harming patients' healthy cells. Furthermore, using the transferred plasma-based configuration, the high-voltage side is kept away to enhance patient safety for future clinical applications. In addition, compared to the large area irradiated with the volume DBD to kill or inactivate melanoma cancer cells [22], the small area of the transferred plasma (with a diameter of around 1~2 mm) are more applicable and allow us to localize the plasma treatment on the skin cancer cells/tumors.

RONS are considered to be bioactive products generated by treatment with CAP. Previous findings show that the conventional CAPs offer an approach for treating melanoma skin cancer by RONS induction of cellular apoptosis [22–24]. Induction of cellular apoptosis by the conventional CAP treatment occurs through the establishment of RONS. The formation of RONS by conventional CAP can also cause damage to DNA and cancer cell death. On the other hand, catalase (CAT) is responsible for neutralizing hydrogen peroxide and for abrogating toxic RONS production [25] possibly by breaking hydrogen peroxide down into water and oxygen. Herein, it is also investigated whether the addition of CAT is able to prevent the toxic effects of RONS, generated by transferred CAP as a developed plasma source.

In this study, an argon-based transferred CAP was developed for melanoma skin cancer cells in vitro. First, the transferred CAP with argon gas was characterized electrically and spectroscopically, then it was utilized for melanoma (B16F10) cell and non-malignant (L929) cell treatments individually, as well as melanoma (B16F10) cell treatment with the addition of CAT.

2. Materials and Methods

2.1. Transfer-CAP System

A scheme of transferring CAP using a silicone tube was utilized to produce a flexible plasma for the cell treatment. The experimental setup of the transferred CAP is illustrated in Figure 1. In this configuration, a 10-cm-long silicone tube was responsible for transferring the plasma jet. Argon gas was utilized in this investigation as it is known to be an affordable gas compared to other noble gases, such as helium and neon. Pure argon gas was adjusted by a mass flow controller to a flow rate of 8 standard liters per minute (SLM).

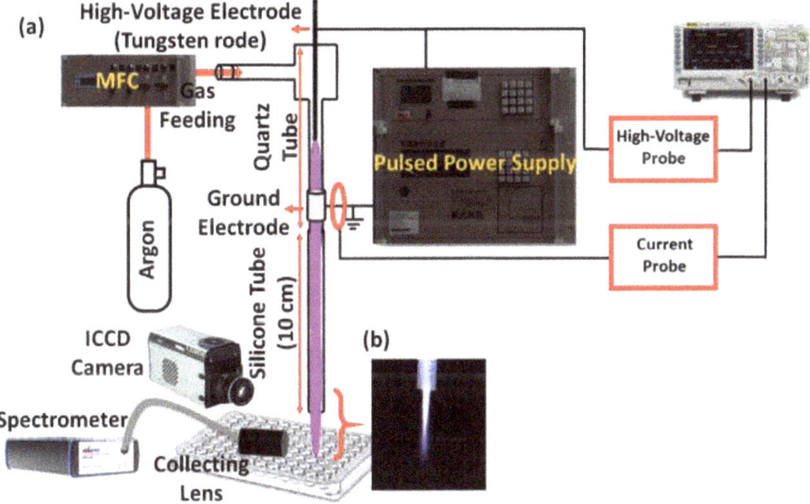

Figure 1. (**a**) Experimental setup of the transferred argon plasma jet system including an ICCD camera for emission characterization and spectrometer for spectroscopic characterization. (**b**) Image of the transferred cold atmospheric plasma (CAP) including argon gas.

2.2. Transfer-CAP Generation

2.2.1. Electrical Measurement

A positive microsecond pulsed discharge with an on-time of 27 μs, an off-time of 47 μs (at a frequency of 12.5 kHz and duty cycle of 36%), and a peak voltage of 8.5 kV provided electrical discharges. A typical voltage-current characteristic corresponding to the transfer-CAP system is shown in Figure 2.

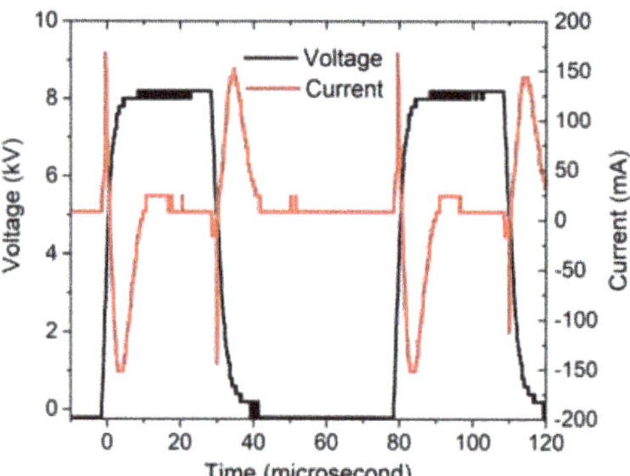

Figure 2. A typical waveform of applied voltage and plasma current.

2.2.2. Emission and Spectroscopic Measurements

An intensified charge-coupled device (ICCD) and optical emission spectroscopy (OES) were utilized to characterize the reactive species generated by the transferred CAP when it interacted with ambient air. The measurement region was adjusted to be downstream of the jet system as shown in Figure 1a.

2.3. Cell Experiments

L929 (a non-malignant mouse fibroblast cell line, accession number: ATCC® CCL-1™) and B16F10 (a mouse melanoma skin cancer cell line, accession number: ATCC® CRL-6475™) cells used for this study were cultured in a cell incubator under standard culture conditions: humidified 5% CO_2, at atmospheric pressure, at 37 °C in Dulbecco's modified Eagle medium with 1% (v/v) penicillin/streptomycin, 10% (v/v) fetal bovine serum (FBS), 1% (v/v) 1-glutamine, and 1% (v/v) non-essential amino acids (NEAAs).

2.3.1. Cell Viability Analysis

We individually seeded L929 and B16F10 cells into 96-well plates (at 10^4 cells/well, with 5% CO_2, at atmospheric pressure, at 37 °C) then cultured them for 2 days until they reached confluence. Before plasma treatment, we removed all culture medium from all wells and treated cells with transferred CAP for 30 s (with and without the addition of CAT from bovine liver with the concentration of 1 mg/50 mL). Subsequently, to test the cell viability after transfer-CAP treatment, we added fresh culture medium (200 µL/well) and a 3-(4,5-dimethylthiazol-2-yl)-2,5-diphenyltetrazolium bromide (MTT) solution (1 mg/mL, 20 µL/well) for a 1.5-h incubation. The supernatant medium was then removed, and dimethyl sulfoxide (DMSO) solution (200 µL/well) was added to dissolve the cellular formazan crystals which had formed. Formazan was quantified with a plate reader (SpectraMax 190) at a wavelength of 570 nm. Data are expressed as an average absorbance (optical density; OD) of triplicate experimental samples + standard deviation (SD) of the average.

2.3.2. Live/Dead Experiment

We seeded B16F10 cells (2×10^5 cells/dish into 35-mm confocal dishes) and cultured them for 2 days in a 5% CO_2 atmosphere at 37 °C. After transfer-CAP treatment, morphological changes in melanoma cells were assessed following staining protocols of a LIVE/DEAD® Viability (Calcein AM)/Cytotoxicity (EthD-III) Assay Kit (ThermoFisher). Fluorescent signals of live/dead cells were then observed with an IX81 optical microscope (Olympus, Tokyo, Japan). Morphological changes of cells treated with transferred CAP were also microscopically analyzed with the IX81 optical microscope (Olympus).

2.3.3. Fluorescence RONS and Catalase Level Analysis in Cells

In this study, B16F10 cells were seeded on confocal dishes incubated under 5% CO_2 at 37 °C until adherence and confluence were achieved. Before CAP treatment, the medium was removed. Then, transferred CAP was applied for 30 s. Fluorescent 4′,6-diamidino-2-phenylindole (DAPI) was used to stain cell nuclei. Cells were stained with 2′,7′-dichlorodihydrofluorescein diacetate (H2DCF-DA) [26] to detect the amount of intracellular RONS generated using an IX81 fluorescent microscope (Olympus) after treatment. The immunofluorescence was employed to examine the expression of catalase in L929 cells and B16F10 cells. The bioactivity of catalase was assessed with fluorescence. Cells were seeded into 96-well (10^4 cells/wells) until attachment and blocked by 1% of serum. For immunofluorescence staining, catalase primary antibodies conjugating fluorescein isothiocyanate diluted 1:39 in PBS were incubated with the cells for 1 h. The unbound antibodies were washed with PBS. Fluorescence emission measurements were determined through a fluorescence microplate reader (n = 13).

2.3.4. Statistical Analysis

Experimental data are presented as the average (AVE) ± SD. Test results were statistically evaluated using Student's t-test. The obtained values were considered to be statistically significant at $p < 0.05$.

3. Results and Discussion

3.1. *Transfer-CAP Characterization*

Figure 3a shows the emission of the transferred CAP. Using optical filters with 50% transmission, which were mounted in front of the ICCD camera, hydroxyl, and atomic oxygen radicals were observed as shown in Figure 3b,c, respectively. The dominant species of the spectra measured by OES were a hydroxyl peak (309 nm) and a nitrogen band (300~450 nm) as well as an argon band, as shown in Figure 3d. A similar spectrum related to the conventional argon-based plasma jet was reported for melanoma tumor treatment [27]. Moreover, a small peak of atomic oxygen radical (777.1 nm) was observed in the spectra (Figure 3e). The small peak of atomic oxygen radical was also reported in a radio-frequency plasma jet with argon [28] and argon/oxygen gas feeds [29]. Although the gas feed used in our system was argon, the generated RONS such as hydroxyl and atomic oxygen radicals as well as nitrogen species observed by OES were mainly respectively generated via energetic electron collisions with water molecules, oxygen, and nitrogen of the ambient air [30]. As a result, the presence of RONS generated by the transferred CAP was confirmed by emission and spectroscopic characterizations.

Gas Temperature

Biomedical applications of CAP first utilized plasma at thermal equilibrium, which was based on thermal energy for tissue removal, disinfection, and cauterization of thermally stable biomedical instruments. However, it is not suitable for heat-sensitive substances such as living tissues, as cells cannot tolerate such high temperatures. Recent advances in novel plasma sources, which can generate transferred CAP under atmospheric pressure in an open space at nearly room temperature, have allowed direct contact between live human cells and plasma yet avoid thermal damage. Implementation of cold plasma is essential for biological or medical treatment (Figure 4a), as only plasma with a temperature slightly higher than the body temperature which can be utilized to avoid harm and distress to patients' cells. As shown in Figure 4b,c, an infrared (IR) camera was used to confirm that the gas temperature remained near room temperature after passing through an extended silicon tube. The thermal image data (Figure 4b,c) show that the temperature in the area of the transferred CAP remained under 30 °C during plasma treatment. For instance, a conventional argon-based CAP with a reported temperature of 35 °C to 40 °C was utilized for melanoma cells treatment [23]. A helium-based CAP with a temperature-controlled environment was also presented for melanoma treatment [24] although using argon is more affordable than helium gas. On the other hand, as mentioned in the introduction, using argon gas leads to a heating problem that causes a relatively high gas temperature compared to the room temperature. The treated biological cells and tissue could be damaged by heating with a temperature around 40 °C although there is no specific threshold temperature for heating damage [31]. In our case, during the plasma treatment, as shown in Figure 4b, the thermal image presents a gas temperature less than 30 °C, which is much safer for avoiding possible heating damage to biological cells and tissue.

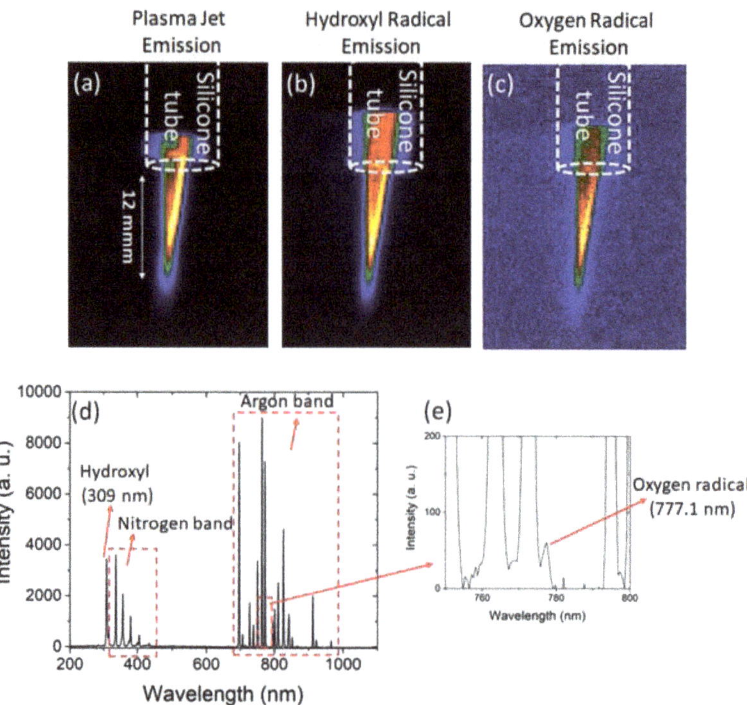

Figure 3. Transferred cold atmospheric plasma (CAP) emission captured by an ICCD camera with an exposure time of 0.5 s. (**a**) Transferred CAP emission without an optical filter. (**b**) Hydroxyl radical emission using an optical filter with 50% transmission. (**c**) Atomic oxygen radical emission using an optical filter with 50% transmission. (**d**) A typical spectrum obtained with a spectrometer during transferred CAP treatment including hydroxyl and (**e**) atomic oxygen radicals.

3.2. Cell Viability

Cell lines of non-malignant murine (L929) fibroblasts and malignant murine melanoma (B16F10) cells were employed in this study. We examined whether the transferred CAP could kill B16F10 cancer cells in an in vitro model as well as explore the potential mechanisms that allow for the specific ablation of cancer cells without affecting non-malignant (L929) cells. We cultured cells in cell medium as shown in Figure 5a. Non-malignant (L929) cells and cancerous B16F10 cells were next collected by trypsinization and centrifugation (at 1200 rpm for 5 min). Collected cells were sub-cultured until used further. For plasma treatment, we removed the culture medium and applied the transfer-CAP system to cells. As a result, after transfer-CAP treatment, using an MTT assay and visible microscopy, we investigated cell viability and morphology as shown in Figure 5b,c, respectively. Cell viability results showed that the number of viable non-malignant (L929) cells was ca. 2~3-fold higher than the number of cancer B16F10 cells after the transferred CAP treatment. Therapeutic effect is one of the most imperative considerations of cancer treatment. Spectroscopic (Figure 3d,e) and MTT (Figure 5b) data suggest that the generation of RONS by transferred CAP preferentially killed malignant melanoma (B16F10) cells without exhibiting significant cytotoxicity toward non-malignant (L929) cells. Similar published findings also supported our experimental data [23,24]. In addition, the optical morphological data showed that the transferred CAP resulted in the separation of cancer cells compared to the before transfer-CAP treatment (Figure 5c). As shown in Figure 5c, optical image data suggests that the transferred CAP also could cause cell shrinkage and damage cancer cells. Transfer-CAP treatment thus led to morphological changes of melanoma cancer cells.

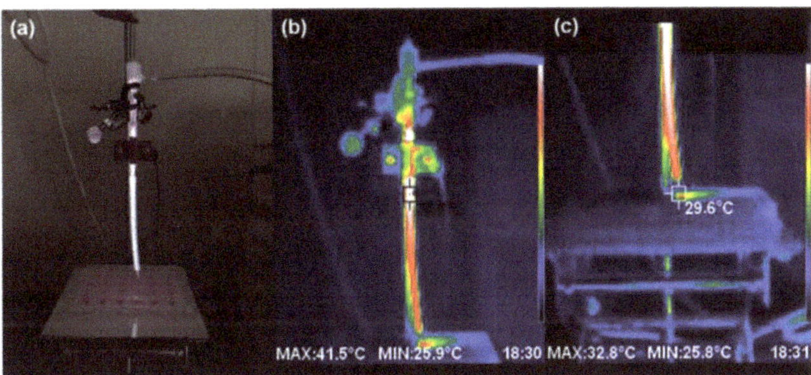

Figure 4. (**a**) The image of transferred cold atmospheric plasma, (**b**,**c**) thermal images captured by the IR camera.

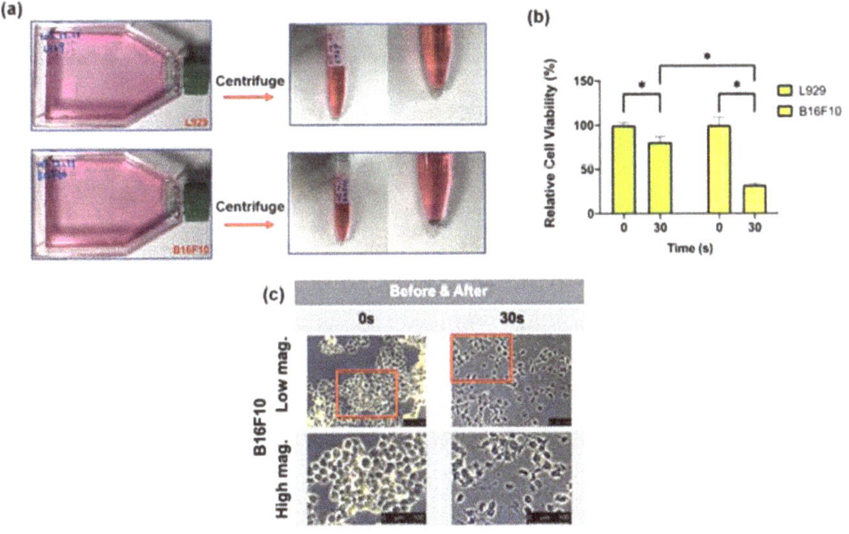

Figure 5. (**a**) Cell culture and collection of L929 and B16F10 cells. (**b**) Cell viability of L929 and B16F10 cells before and after plasma treatment. (**c**) B16F10 cell morphology was imaged by optical microscopy at different magnifications before and after transferred CAP treatment. (* $p < 0.05$).

3.3. Live/Dead Experiment (Control, 30 s, and 30 s with CAT)

The transferred CAP generates RONS, including hydroxyl, atomic oxygen radicals, and nitrogen species, as shown in Figure 3. RONS-induced death of cells is a strategy for cancer treatment [32]. At low levels, RONS play crucial roles during redox homeostasis and in biological signals. However, high levels of RONS are able to break the balance, causing irreversible oxidative damage to lipids, carbohydrates, proteins, and DNA. RONS can be catalytically decomposed into water and oxygen by CAT, which defends cells from RONS oxidative damage [32]. CAT is a biological enzyme that plays key roles in cellular antioxidant defense mechanisms precipitated by the accumulation of RONS. To further examine the CAT and CAP relationship on B16F10 cancer cells, CAT was added to melanoma cancer (B16F10) cells before transfer-CAP treatment in the following cell viability test. In the live/dead study, 1 mg CAT/50 mL was pre-added to the cell culture medium for the co-culture of B16F10 cancer cells, and then cells were treated with transferred CAP. As shown in Figure 6a, the fluorescent-green represents live cells and red represents dead cells, which differ in the two treated groups. In the control group (B16F10 cells plus fluorescent dyes), 30 s of CAP treatment, leads to cell death. In the group after co-culture with CAT, it can clearly be observed that CAT might preserve cells alive even after treatment with transferred CAP. In addition, the MTT assay was used for further quantification (Figure 6b). The MTT data suggest that CAT adds potential protection to the transferred CAP resulting in a higher cell viability, compared to cells treated by the transferred CAP without CAT (Figure 6b). Furthermore, Image J software was used to quantitatively calculate green live cell fluorescent signals (Figure 6c). As shown in Figure 6b,c, the transferred CAP-treated group given CAT had higher cell viability than that of the transferred CAP-treated group without CAT. The addition of biological CAT is recognized to be a mechanism of cellular defense against RONS. Its function helps to balance the amount of cell RONS, whereas an imbalance between RONS and RONS-scavenging enzymes can lead to an event called oxidative stress [33].

3.4. Cell RONS Analysis (Control, 30 s, and 30 s with CAT)

Cellular RONS were seen to increase under CAP treatment, as shown in Figure 7a. The production of RONS in cells exists in equilibrium with antioxidant defenses. At moderate levels, RONS are believed to be important for regulating physiological and biological functions involved in development, including cell-cycle progression and differentiation, proliferation, and migration. RONS play vital roles in the immune system, maintaining a redox balance, and are associated with the bioactivation of various cell signaling pathways. Excess RONS in cells causes damage to cellular proteins, lipids, nucleic acids, organelles, and membranes that can lead to activation of cell death processes, including cell apoptosis. Apoptosis is a greatly modulated process that is critical for the survival and development of multicellular organisms. These multicellular organisms usually must discard cells that are possibly harmful or superfluous or that possess accumulated mutations. Apoptosis features a representative set of biochemical, pathological, and morphological aspects whereby cells undergo a sequence of self-destruction [34].

We used a cellular RONS analytical method to confirm the CAT and CAP roles on B16F10 cancer cells. Cell RONS levels can be determined in live cells by converting the non-fluorescent 2′,7′-dichlorofluorescein diacetate (DCFDA) that is oxidized into the fluorescent tracer, 2′, 7′-dichlorofluorescein (DCF). The generated fluorescent signal is directly proportional to the amount of DCFDA chemically oxidized into DCF. As the fluorescent emission is at the wavelength of ca. 529 nm, it can be determined by fluorescence microscopy, thereby measuring hydroxyl, peroxyl, and other RONS bioactivity in tested cells. Thus, melanoma cells were treated with the transferred CAP in the same way as the aforementioned cell culture method, then cells were stained with DAPI, and DCFDA RONS fluorescence was detected. The fluorescence intensity of RONS was much higher in CAP-treated cells (Figure 7a). The generated cellular RONS signals potentially came from the contributions of applied CAP or cell death after CAP treatment. However, the detailed mechanisms of RONS generation inside cells need to be investigated in the future. At the same time, the group that received CAT, after quantitative plasma treatment and analysis with Image J software (Figure 7b), possessed lower RONS fluorescent signals than the transfer-CAP-treated group without CAT, which confirms the original hypothesis.

Furthermore, an interesting research [35], reported that CAP can help in wound healing (non-malignant L929 cells). According to published literature, a moderate ROS level is vital for promoting cell proliferation [36]. A moderate level of ROS plays an essential role in the cell signaling that controls cell survival and cell proliferation. However, a rise in ROS level could damage cell components for example DNA, lipids, proteins, triggering an imbalance among cellular reduction-oxidation (redox) situations and causing homeostasis disruption (in the case of cold-plasma treated cancer cells). By comparing the intensities of the immune-stained CAT level, the staining protein level detected by fluorescence microplate reader in non-malignant L929 cells group was much higher than that in the malignant B16F10 cells cell lines ($p < 0.05$), which suggests possible higher expression of CAT in L929 cells. The higher CAT expression from L929 cells group might possess an augmented ROS scavenging efficacy, compared to that of the B16F10 cells group, thus causing an anticancer effect when CAP is applied on B16F10 cells (Figure 7c). At present, thus we conclude that CAP can kill B16F10 melanoma cells effectively.

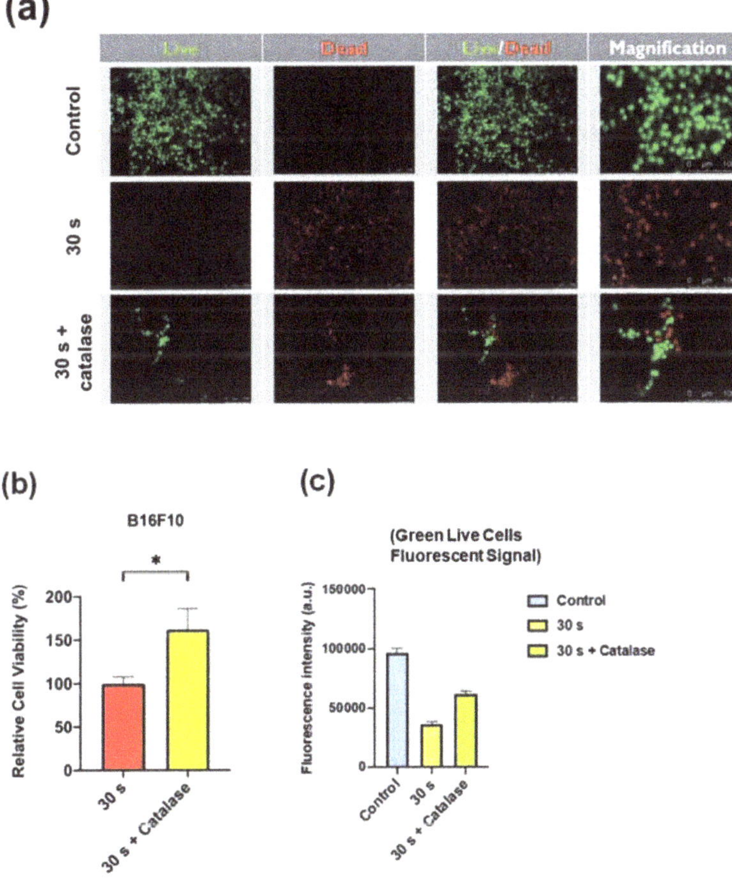

Figure 6. (**a**) Fluorescence image of B16F10 cells treated with transferred cold atmospheric plasma (CAP) after 30 s without and with CAT. Green fluorescence represents live cells, while red fluorescence represents dead cells. (**b**) Cell viability (MTT) after 30 s of treatment without and with CAT. Cell viability was calculated with normalizing the survival of the treated cells group (transferred CAP 30 s) as 100%. (**c**) Fluorescence intensity after 30 s of treatment without and with CAT (green alive cell fluorescent signals). (* $p < 0.05$) Control: B16F10 cells plus fluorescent dyes.

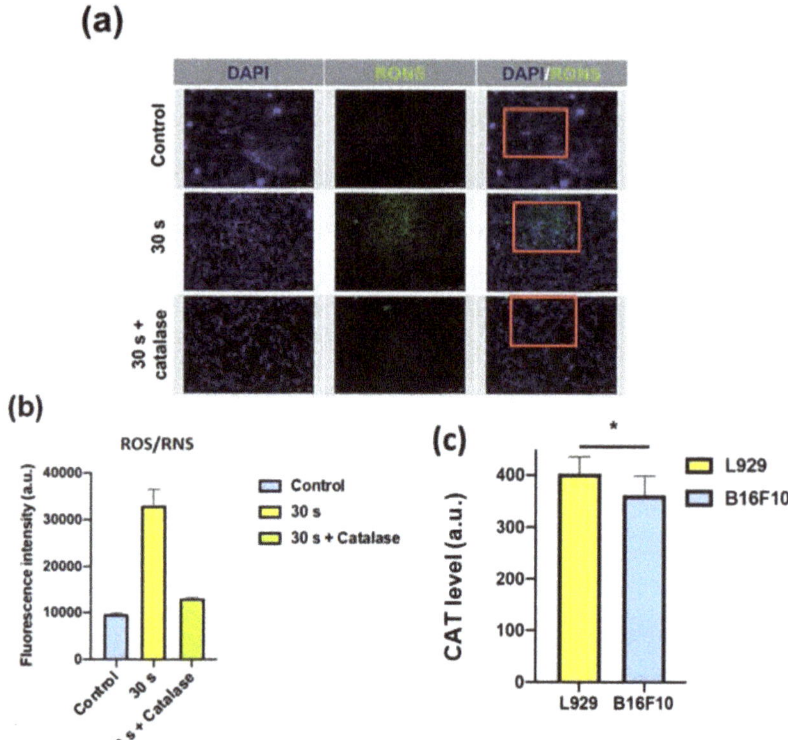

Figure 7. Cellular (B16F10 cells) reactive oxygen and nitrogen species (RONS) generation. (**a**) Qualitative and (**b**) quantitative RONS data indicating that cells that received catalase (CAT) could inhibit RONS generation after cold atmospheric plasma (CAP) treatment compared to the CAP-treated group (without CAT). (**c**) Cellular CAT levels were determined by immunofluorescence. (* $p < 0.05$).

4. Conclusions

A transferred argon-driven plasma jet using a silicone tube was employed to make a flexible plasma jet system for melanoma skin cancer cell treatment. Results showed that the transferred plasma jet was able to effectively kill (inactivate) melanoma skin cancer cells. In addition, the presence of the CAT enzyme confirmed that RONS play key roles in killing melanoma cancer cells. The transferred CAP treatment was effective with a short time duration of 30 s, whereas non-malignant (L929) cells were less damaged during plasma treatment. This means that the transferred plasma jet can be considered for melanoma skin cancer therapy. Even though our and other groups proved CAP has anticancer outcomes, detailed investigation is still needed to understand the mechanisms underlying this result.

Author Contributions: Investigation, data analysis, Y.-H.C.; supervision, J.-H.H.; investigation, data analysis, I.-T.W.; investigation, P.-R.J., Y.-Y.Y.; supervision, J.-W.L.; writing—original draft preparation, writing—review and editing, N.B., E.-Y.C. All authors have read and agreed to the published version of the manuscript.

Funding: This research work was financially supported by the Ministry of Science and Technology, Taiwan (MOST 108-2320-B-038-061-MY3 and 108-2221-E-038-017-MY3).

Institutional Review Board Statement: Not applicable.

Informed Consent Statement: Not applicable.

Data Availability Statement: Not applicable.

Acknowledgments: Not applicable.

Conflicts of Interest: The authors declare no conflict of interest.

References

1. Erdei, E.; Torres, S.M. A new understanding in the epidemiology of melanoma. *Expert Rev. Anticancer. Ther.* **2010**, *10*, 1811–1823. [CrossRef] [PubMed]
2. D'Orazio, J.; Jarrett, S.; Amaro-Ortiz, A.; Scott, T. UV radiation and the skin. *Int. J. Mol. Sci.* **2013**, *1*, 12222–12248. [CrossRef] [PubMed]
3. Ikegawa, S.; Saida, T.; Obayashi, H.; Sasaki, A.; Esumi, H.; Ikeda, S.; Kiyohara, Y.; Hayasaka, K.; Ishihara, K. Cisplatin combination chemotherapy in squamous cell carcinoma and adenoid cystic carcinoma of the skin. *J. Dermatol.* **1989**, *16*, 227–230. [CrossRef] [PubMed]
4. Kalal, B.S.; Upadhya, D.; Pai, V.R. Chemotherapy resistance mechanisms in advanced skin cancer. *Oncol. Rev.* **2017**, *11*, 326. [CrossRef]
5. Abdullah, Z.; Mohtar, J.A.; Zaaba, S.K. Review on Melanoma Skin Cancer Treatment by Cold Atmospheric Plasma. *J. Telecommun. Electron. Comput. Eng. JTEC* **2018**, *10*, 97–100.
6. Baskar, R.; Lee, K.A.; Yeo, R.; Yeoh, K.-W. Cancer and radiation therapy: Current advances and future directions. *Int. J. Med. Sci.* **2012**, *9*, 193. [CrossRef]
7. Burm, K. Plasma: The fourth state of matter. *Plasma Chem. Plasma Process.* **2012**, *32*, 401–407. [CrossRef]
8. Laroussi, M. Cold plasma in medicine and healthcare: The new frontier in low temperature plasma applications. *Front. Phys.* **2020**, *8*, 74. [CrossRef]
9. Duarte, S.; Panariello, B.H. Comprehensive biomedical applications of low temperature plasmas. *Arch. Biochem. Biophys.* **2020**, *693*, 108560. [CrossRef] [PubMed]
10. Martines, E. *Special issue "Plasma Technology for Biomedical Applications"*; Multidisciplinary Digital Publishing Institute: Basel, Switzerland, 2020.
11. Privat-Maldonado, A.; Schmidt, A.; Lin, A.; Weltmann, K.-D.; Wende, K.; Bogaerts, A.; Bekeschus, S. Ros from physical plasmas: Redox chemistry for biomedical therapy. *Oxidative Med. Cell. Longev.* **2019**, *2019*. [CrossRef]
12. Borges, A.C.; Kostov, K.G.; Pessoa, R.S.; de Abreu, G.M.A.; Lima, G.d.G.; Figueira, L.W.; Koga-Ito, C.Y. Applications of Cold Atmospheric Pressure Plasma in Dentistry. *Appl. Sci.* **2021**, *11*, 1975. [CrossRef]
13. Privat-Maldonado, A.; Bogaerts, A. *Plasma in Cancer Treatment*; Multidisciplinary Digital Publishing Institute: Basel, Switzerland, 2020.
14. Stope, M.B. Plasma oncology-Physical plasma as innovative tumor therapy. *J. Cancer* **2020**, *53*, 56.
15. Keidar, M.; Shashurin, A.; Volotskova, O.; Stepp, M.A.; Srinivasan, P.; Sandler, A.; Trinket, B. Cold atmospheric plasma in cancer therapy. *Phys. Plasmas* **2013**, *20*, 057101. [CrossRef]
16. Brandenburg, R. Dielectric barrier discharges: Progress on plasma sources and on the understanding of regimes and single filaments. *Plasma Sources Sci. Technol.* **2017**, *26*, 053001. [CrossRef]
17. Robert, E.; Vandamme, M.; Brullé, L.; Lerondel, S.; le Papebe, A.; Sarrona, V.; Rièsa, D.; Darnya, T.; Doziasa, S.; Colletaf, G.; et al. Perspectives of endoscopic plasma applications. *Clin. Plasma Med.* **2013**, *1*, 8–16. [CrossRef]
18. Onyshchenko, I.; de Geyter, N.; Nikiforov, A.Y.; Morent, R. Atmospheric pressure plasma penetration inside flexible polymeric tubes. *Plasma Process. Polym.* **2015**, *12*, 271–284. [CrossRef]
19. Prysiazhnyi, V.; Saturnino, V.F.; Kostov, K.G. Transferred plasma jet as a tool to improve the wettability of inner surfaces of polymer tubes. *Int. J. Polym. Anal. Charact.* **2017**, *22*, 215–221. [CrossRef]
20. Schweigert, I.; Zakrevsky, D.; Gugin, P.; Yelak, E.; Golubitskaya, E.; Troitskaya, O.; Koval, O. Interaction of cold atmospheric argon and helium plasma jets with bio-target with grounded substrate beneath. *Appl. Sci.* **2019**, *9*, 4528. [CrossRef]
21. Wang, S.; der Gathen, V.S.; Döbele, H. Discharge comparison of nonequilibrium atmospheric pressure Ar/O_2 and He/O_2 plasma jets. *Appl. Phys. Lett.* **2003**, *83*, 3272–3274. [CrossRef]
22. Fridman, G.; Shereshevsky, A.; Jost, M.M.; Brooks, A.D.; Fridman, A.; Gutsol, A.; Vasilets, V.; Friedman, G. Floating electrode dielectric barrier discharge plasma in air promoting apoptotic behavior in melanoma skin cancer cell lines. *Plasma Chem. Plasma Process.* **2007**, *27*, 163–176. [CrossRef]
23. Alimohammadi, M.; Golpour, M.; Sohbatzadeh, F.; Hadavi, S.; Bekeschus, S.; Niaki, H.A.; Valadan, R.; Rafiei, A. Cold atmospheric plasma is a potent tool to improve chemotherapy in melanoma in vitro and in vivo. *Biomolecules* **2020**, *10*, 1011. [CrossRef]
24. Zucker, S.N.; Zirnheld, J.; Bagati, A.; DiSanto, T.M.; des Soye, B.; Wawrzyniak, J.A.; Etemadi, K.; Nikiforov, M.; Berezney, R. Preferential induction of apoptotic cell death in melanoma cells as compared with normal keratinocytes using a non-thermal plasma torch. *Cancer Biol. Ther.* **2012**, *13*, 1299–1306. [CrossRef] [PubMed]
25. Cloux, A.-J.; Aubry, D.; Heulot, M.; Widmann, C.; ElMokh, O.; Piacente, F.; Cea, M.; Nencioni, A.; Bellotti, A.; Bouzourène, K.; et al. Reactive oxygen/nitrogen species contribute substantially to the antileukemia effect of APO866, a NAD lowering agent. *Oncotarget* **2019**, *10*, 6723. [CrossRef]
26. Chwa, M.; Atilano, S.R.; Reddy, V.; Jordan, N.; Kim, D.W.; Kenney, M.C. Increased stress-induced generation of reactive oxygen species and apoptosis in human keratoconus fibroblasts. *Investig. Ophthalmol. Vis. Sci.* **2006**, *47*, 1902–1910. [CrossRef]

27. Rafiei, A.; Sohbatzadeh, F.; Hadavi, S.; Bekeschus, S.; Alimohammadi, M.; Valadan, R. Inhibition of murine melanoma tumor growth in vitro and in vivo using an argon-based plasma jet. *Clin. Plasma Med.* **2020**, *19*, 100102. [CrossRef]
28. Cullen, P.; Milosavljević, V. Spectroscopic characterization of a radio-frequency argon plasma jet discharge in ambient air. *Prog. Theor. Exp. Phys.* **2015**, *2015*. [CrossRef]
29. Lim, J.-P.; Uhm, H.S.; Li, S.-Z. Influence of oxygen in atmospheric-pressure argon plasma jet on sterilization of *Bacillus atrophaeous* spores. *Phys. Plasmas* **2007**, *14*, 093504. [CrossRef]
30. Lietz, M.A.; Kushner, M.J. Molecular admixtures and impurities in atmospheric pressure plasma jets. *J. Appl. Phys.* **2018**, *124*, 153303. [CrossRef]
31. Yarmolenko, P.S.; Moon, E.J.; Landon, C.; Manzoor, A.; Hochman, D.W.; Viglianti, B.L.; Dewhirst, M.W. Thresholds for thermal damage to normal tissues: An update. *Int. J. Hyperth.* **2011**, *27*, 320–343. [CrossRef] [PubMed]
32. Sun, K.; Gao, Z.; Zhang, Y.; Wu, H.; You, C.; Wang, S.; An, P.; Suna, C.; Sun, B. Enhanced highly toxic reactive oxygen species levels from iron oxide core–shell mesoporous silica nanocarrier-mediated Fenton reactions for cancer therapy. *J. Mater. Chem. B* **2018**, *6*, 5876–5887. [CrossRef]
33. Nantapong, N.; Murata, R.; Trakulnaleamsai, S.; Kataoka, N.; Yakushi, T.; Matsushita, K. The effect of reactive oxygen species (ROS) and ROS-scavenging enzymes, superoxide dismutase and catalase, on the thermotolerant ability of Corynebacterium glutamicum. *Appl. Microbiol. Biotechnol.* **2019**, *103*, 5355–5366. [CrossRef] [PubMed]
34. Redza-Dutordoir, M.A.; Averill-Bates, D. Activation of apoptosis signalling pathways by reactive oxygen species. *Biochim. Biophys. Acta BBA Mol. Cell Res.* **2016**, *1863*, 2977–2992. [CrossRef] [PubMed]
35. Liu, J.-R.; Xu, G.-M.; Shi, X.-M.; Zhang, G.-J. Low temperature plasma promoting fibroblast proliferation by activating the NF-κB pathway and increasing cyclinD1 expression. *Sci. Rep.* **2017**, *7*, 11698. [CrossRef] [PubMed]
36. Kim, S.J.; Kim, H.S.; Seo, Y.R. Understanding of ROS-Inducing Strategy in Anticancer Therapy. *Oxidative Med. Cell. Longev.* **2019**, *2019*, 5381692. [CrossRef] [PubMed]

MDPI

Article

Vortex Breakdown Control by the Plasma Swirl Injector

Gang Li [1,2,*], Xi Jiang [3], Wei Du [1,2], Jinhu Yang [1,2,*], Cunxi Liu [1,2], Yong Mu [1,2] and Gang Xu [1,2]

[1] Key Laboratory of Light Duty Gas Turbine, Institute of Engineering Thermophysics, Chinese Academy of Sciences, Beijing 100190, China; duwei@iet.cn (W.D.); liucunxi@iet.cn (C.L.); muyong@iet.cn (Y.M.); xug@iet.cn (G.X.)
[2] School of Aeronautics and Astronautics, University of Chinese Academy of Sciences, Beijing 100049, China
[3] School of Engineering and Materials Science, Queen Mary University of London, Mile End Road, London E1 4NS, UK; xi.jiang@qmul.ac.uk
* Correspondence: ligang@iet.cn (G.L.); yangjinhu@iet.cn (J.Y.)

Abstract: Vortex breakdown, observed in swirling flows, is an interesting physical phenomenon relevant to a wide range of engineering applications, including aerodynamics and combustion. The concept of using a plasma swirler to control vortex breakdown was proposed and tested in this study. The effect of plasma actuation on controlling the onset and development of the vortex breakdown was captured by particle image velocimetry. Flowfield measurement results suggested that, by varying the strength of the plasma actuation, the location and size of the vortex breakdown region was controlled effectively. The plasma swirl injector offers a method for optimal control and efficient utilization of vortex breakdown. The method being proposed here may represent an attractive way of controlling vortex breakdown using a small amount of energy input, without a moving or intrusive part.

Keywords: vortex break down; plasma swirl injector; dielectric barrier discharge; swirling flow control

Citation: Li, G.; Jiang, X.; Du, W.; Yang, J.; Liu, C.; Mu, Y.; Xu, G. Vortex Breakdown Control by the Plasma Swirl Injector. *Appl. Sci.* **2021**, *11*, 5537. https://doi.org/10.3390/app11125537

Academic Editor: Mariusz Jasiński

Received: 7 May 2021
Accepted: 7 June 2021
Published: 15 June 2021

Publisher's Note: MDPI stays neutral with regard to jurisdictional claims in published maps and institutional affiliations.

1. Introduction

As an active flow and combustion control device, the dielectric barrier discharge (DBD) actuator has drawn much attention for its fast response, low power consumption, and simple structure, as described by Roth et al. [1]. The operation of the DBD actuator is purely electric without any moving part, which is attractive to many applications. Recent investigations on plasma actuators have been reviewed by Moreau [2], Corke et al. [3], Wang et al. [4], Kriegseis et al. [5], Leonov et al. [6] and Konstantinidis [7], to name but a few. Flow separation control by the DBD actuator was widely researched. Roupassov et al. [8] observed that the heat released by nanosecond pulse actuation can produce shock waves, where the associated secondary vortex flows disturbed the main flow and caused an efficient transversal momentum transfer into the boundary layer. Flow separation control was investigated experimentally by Little et al. [9] on an airfoil leading edge up to 62 m/s with nanosecond pulse plasma actuator. They pointed out that the plasma actuator was similar to an active trip, which can generate coherent spanwise vortices at post stall. Fujii [10] showed that actuation in burst mode was very effective for controlling flow separation at a Reynolds number of 6.3×10^4, where the three features of the flow structure, associated with flow separation control, were emphasized and guidelines for the effective use of DBD actuators were proposed. Meanwhile, the modelling approach of Shyy et al. [11] was adopted by Hasan et al. [12] to simulate the 3D separated flow over a hump model with the inlet velocity of 34.6 m/s. They demonstrated that the flow separation was completely suppressed with the actuator placed just downstream of the separation point at an applied frequency of 5 kHz. Through flowfield measurement by particle image velocimetry (PIV) and a pressure cap over a wing section, Skourides et al. [13] emphasized that the actuation frequency was critical to the control authority of a nanosecond-DBD actuator. Pescini et al. [14] used a micro plasma actuator to suppress flow separation in the low-pressure turbine of small engines. They pointed out that the sinus waveforms outperformed other waveforms slightly. Lo et al. [15] tested the effective flow separation control over the rear end of a lorry (1:20 scale) by the plasma actuation with the free stream velocity at 30 m/s. They found that the shear layer was deflected more downwards due to the actuation. However, the flow characteristics in the wake region was not significantly modified by the actuation and they concluded that no flow control effect was observed in their study. All of these studies demonstrate the wide range of applications of plasma actuation, with further potentials to be explored.

Using ionic wind (which is the airflow induced by electrostatic forces linked to plasma discharge) for local cooling is also an interesting research topic. Go et al. [16] demonstrated that heat transfer was increased by ionic wind through distortion of the boundary layer. Roy and Wang [17] developed the idea of film cooling enhancement by plasma actuator. Through PIV and infrared thermography measurements, Audier et al. [18] found that the actuation deflected the jet toward the wall and delayed its diffusion into the cross flow, and, as a result, the effectiveness of the film cooling was increased. Through numerical simulation, the mechanism of film cooling improvement by plasma actuator was analyzed by Xiao et al. [19]. They found that the counter rotating vortex pairs were weakened by the actuation, which led to less interaction and reduced mixing between the main flow and the jet flow. A pressure sensitive paint technique was adopted by Kim et al. [20] to investigate the effect of DBD actuation on the film cooling effectiveness of a 30° slot with the mainstream velocity at 10 m/s. They pointed out that the improvement was not significant and that the actuator configuration should be optimized. Uehara and Takana [21] developed a plasma actuator cooling device in millimeter scale channels with a height of 2.5 mm, 10 mm, and 50 mm, without an upper wall. They found that the heat transfer coefficient was inversely proportional to the channel height. However, when the height was 2.5 mm, the heat transfer coefficient was larger than that of the height of 5 mm. They attributed this to the suppression of backward flow which impeded the cooling performance. They were optimistic about the feasibility of the plasma actuator cooling for a confined space. The plasma actuator was also adopted to control noise and vibration. The experiments of Hebrero et al. [22] showed that plasma actuators exhibited the ability to suppress vortex-induced vibration around a rigid circular cylinder. During their experiments the flow velocity was varied between 3 to 4.25 m/s ($6000 < Re < 8500$). Yokoyama et al. [23] demonstrated that the plasma actuation was effective in reducing the cavity tone with the acoustic resonances at freestream velocity of 30 m/s by introducing streamwise vortices with fine displacement. A new corner-type actuator configuration was designed by Jong et al. [24] to control flow induced noise in a cavity. They demonstrated that the aero-acoustic lock-on was suppressed for freestream velocities below 12.5 m/s by the inward-inducing actuator. PIV measurement showed that a secondary circulating flow region was created by the actuation which prevented lock-on. The experimental research of Silva et al. [25] demonstrated that the DBD actuators were able to reduce aerodynamic noise from slats with wind tunnel velocity of 27.4 m/s. They also pointed out that the control authority of the actuators tested was still poor for real flight conditions, and that further research is needed to produce more efficient plasma actuators. The control effects of the plasma actuation on the flow around an oscillating plate was investigated by Sato et al. [26]. They pointed out that it was possible for the DBD actuator to control the lift with a mechanism similar to that of an insect. They found that the appropriate adjustment of the driving time played an important role in lift enhancement. Motta et al. [27] numerically assessed the effectiveness of plasma actuators for load alleviation on a compressor cascade. They demonstrated that it was effective to alter the blade loading by proper triggering of the pressure and suction side of the actuation. A number of studies have been carried out to enhance the performance of DBD actuators, such as the shape of the serpentine actuator [28], wire type electrode [29], microfabricated DBD actuator [30], sawtooth electrode [31], and DBD active grid [32].

Vortex breakdown (VB) observed in swirling flows is an interesting physical phenomenon relevant to a wide range of engineering applications such as aerodynamics (delta wing, and blade tip leakage flow) and combustion. Peckham and Atkinson [33] first reported vortex breakdown on a delta wing at a high angle of attack. The vortex lines near the edges of the wing experience a sudden change in shape when the angle of attack is increased. This phenomenon is referred to as vortex breakdown. Sarpkaya [34] observed three types of vortex breakdown: double helix, spiral, and axisymmetric (bubble). Bubble breakdown is characterized by a stagnation point on the swirl axis, followed by an abrupt expansion of the centerline to form an envelope of recirculating fluid [35]. Leibovich [36] defined vortex breakdown as "a disturbance characterized by the formation of an internal stagnation point on the vortex axis, followed by a reversed flow in a region of limited axial extension". Several theories were proposed to explain the process of vortex breakdown, such as critical state or wave phenomena theory [37], boundary layer separation or flow stagnation [38], and hydrodynamic instability [39]. Shtern [40] recommended the swirl-decay mechanism (SDM) as a simple physical reason for the vortex breakdown occurrence, and also pointed out that SDM indicated efficient means for vortex breakdown control. The flow and boundary conditions have a strong effect on the breakdown process. Sarpkaya [34] observed a periodic transition between bubble-type and spiral-type depending on the flow and boundary conditions. Aithaus et al. [41] observed the transition from bubble-type and spiral-type, both in experiments and numerical simulations, and proposed a feedback model for the initiation and development of the bubble-type breakdown. It is of practical important to study methods of controlling VB, so that it can be enhanced when it is beneficial and weakened when it is detrimental. The main techniques employed to control VB include temperature gradient (Herrada and Shtern [42]), shape modification (Srigrarom and Kurosaka [43]), blowing (Schmucker and Gersten [44], Gutmark and Guillot [45]), rotation of the end walls (Mununga et al. [46]), the addition of near-axis swirl (Husain et al. [47]), and energy deposition (Zheltovodov et al. [48]). Mitchell and Delery reviewed the early research on VB control in detail [49], while the developments in this area are continuously evolving.

Although vortex breakdown should be suppressed on a delta wing, it is commonly adopted in combustion and acts as a stabilizer to enhance reactant mixing and stabilization of the flame by recirculating hot gases into its base. Inspired by the idea of accelerating the same fluid particles continuously, we designed a plasma swirler with the electrode placed in the streamwise direction [50]. The plasma swirl injector is effective in diffusion and premixed flame control [51,52], and can be integrated into a low swirl injector to control the flame lift off height accurately [53,54]. The electrode of the plasma swirler was optimized as a helical shape to enhance its performance [55]. In this study, the concept of using the plasma swirler to control vortex breakdown was proposed and tested. A plasma swirler with a helical shape was adopted to control the vortex breakdown. The plasma actuation, in affecting the onset and development of the vortex breakdown, was captured and analyzed.

2. Layout of the Experimental Setup

The plasma swirl injector with a helical-shaped electrode was adopted for this research. Figure 1 shows the sketch and the photograph of the plasma swirl injector used in the experiment. Design parameters and advantages of the helical shape electrode were discussed previously [55]. The length and inner diameter of the quartz glass tube are 100 mm and 36 mm, respectively. In this study, the actuator was further optimized. The electrodes were arranged perpendicular to the blades at the inlet of the electrode. Compared to that of 170 mm in the earlier study [55], the helix pitch was shortened to 100 mm, which means the electrode was longer and more plasma could be generated at the same voltage. The length of the electrode was approximately 130 mm. The activated and grounded electrodes were stuck to the inside and outside wall of the injector with a width of 2.5 mm and 7.5 mm, respectively.

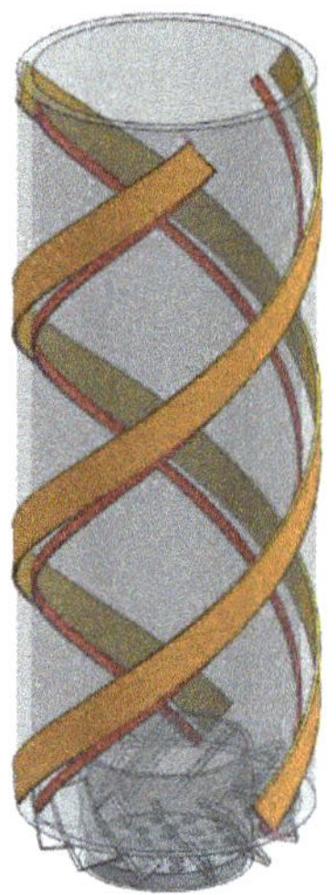

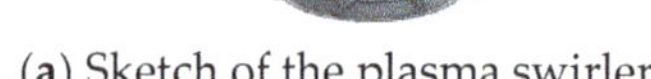

(a) Sketch of the plasma swirler

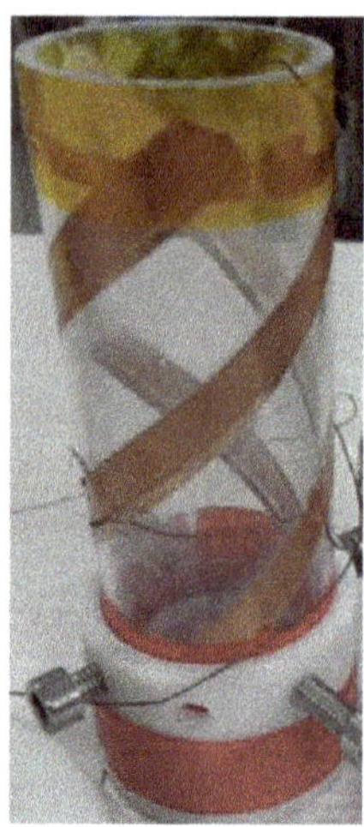

(b) Photograph of the plasma swirler

Figure 1. Sketch and photograph of plasma swirler with helical shape electrodes.

In the experiment, air volumetric flow was 175 l/min and the corresponding bulk flow velocity in the injector was approximately 3 m/s. The power inputs of the plasma generator were approximately 31~76 W and the electrodes on the inner wall were activated with a peak-to-peak amplitude of 12~21 kV. Parameters of the five test cases are listed in Table 1.

Table 1. The five test cases and their experimental conditions.

Test Case	Actuator Status	Waveform Type	Voltage Amplitude (kV)	Voltage Frequency (kHz)	Power Inputs (W)
1	OFF	-	-	-	-
2	ON	Sinusoidal	12	9	31
3	ON	Sinusoidal	15	9	45
4	ON	Sinusoidal	18	9	60
5	ON	Sinusoidal	21	9	76

Figure 2 shows the schematic of the experimental setup in this research. The flowfields of the five test cases were captured by 2D PIV, manufactured by Lavision (Gottingen, Genmany) [55]. A screw compressor supplied compressed air, which was then settled into two tanks of 1.2 m^3. A small wind tunnel, manufactured by 3D printing, was installed on a displacement device. Through the bottom inlet of the wind tunnel, air flowed into the settling chamber, in which two layers of honeycomb were used to improve flow uniformity. A contraction section, with an area ratio of 6:1, was used to further improve the flow quality. A mass flowmeter was utilized to regulate and measure the flow rate.

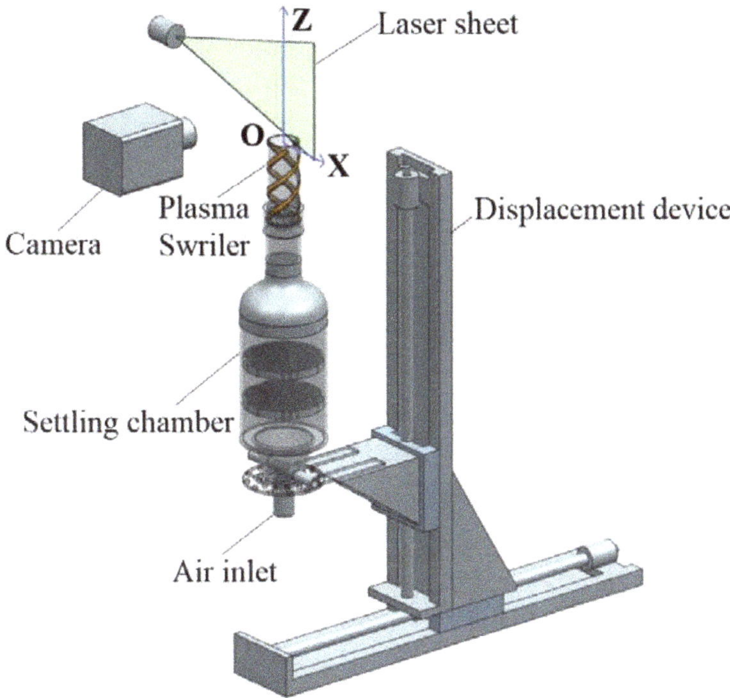

Figure 2. Schematic of the experimental setup.

3. Results and Discussion

The flowfield, obtained by 2D PIV, provides information about the presence and location of vortex breakdown. The leading edge of the VB could be identified by a stagnation point along the centerline. Contours of axial velocity in the centerline plane (the laser plane in Figure 2) were captured for each case to provide a global quantitative and qualitative depiction of the flowfield. Figure 3a–e, from 2D PIV measurements, show the axial velocity magnitude contours and vectors in the XZ plane for the five conditions tested, with a bulk velocity of 3 m/s and a corresponding Reynolds number of 7171 based on the injector diameter. The approximate exit of the injector was at $Z = -2$ mm. Location of a breakdown zone can be identified by the negative axial velocity and recirculating flow.

Generally, the axial velocity distribution of a low swirl flowfield accords with the normal distribution of this type of jet flow, but the maximum axial velocity moves away from the axis due to the effect of the mesh plate of the low swirl injector, forming a double peak velocity profile as shown in Figure 3a. The velocity in the annular region is higher because of the significant downward momentum of the flow in this region, compared with the flow in the central region. The central low velocity region, as well as the shear layer between the non-swirling core and swirling annulus, are critical for flame stabilization. When the swirl jet flows out of the low swirl injector, the low pressure in the center region of the swirl flow recovers gradually, which leads to the generation of the reverse pressure gradient in the axial direction. Without plasma actuation, the axial pressure gradient is not large enough to cause reverse flow. At 12 kV, the decrease of axial velocity near the central region can be observed, where the central low velocity region swells slightly and the degree of flow expansion increases as well, but the actuation is not strong enough to cause flow reversal and no vortex breakdown is observed, as shown in Figure 3b. Increasing the strength of actuation to 15 kV resulted in the generation of a stagnation point as a result of the reverse pressure gradient increase and the decrease of the axial velocity. Although the low velocity region and the degree of flow expansion remained almost unchanged, flow reversal occurred in a small region and the onset of the breakdown bubble was observed in regions close to the centerline at around Z = 15 mm, as shown in Figure 3c. Bubble-type breakdown is characterized by a nearly axisymmetric region of reversed flow, with a stagnation point at the forward end. As the actuation voltage was further increased to 18 kV, the stable bubble-type vortex breakdown was observed to migrate upstream and was fully established with a forward stagnation point at about Z = 1 mm and a backward stagnation point at Z = 36 mm, as shown in Figure 3d. Its outer shape was nearly axisymmetric and can be visualized as a bubble. Due to the existence of the vortex, the outer boundary of the jet expanded immediately after it flowed out of the injector. As the strength of actuation was increased to 21 kV, the breakdown zone grew further due to the enhanced backflow, and its nose becomes less pointed and wider, as shown in Figure 3e. The breakdown zone shifted upstream with the forward stagnation point penetrating into the injector, but the position of the backward stagnation point changed very slightly compared to that of 18 kV. This shows that the actuation does not significantly affect the overall location of the bubble, although it enhances its size. The interior structures of the bubble were nearly the same for the two cases.

Based on these results, it can be concluded that the 15 kV actuation is strong enough to trigger the onset of vortex breakdown with a relatively low power input. The upstream stagnation point and the center of the breakdown bubble migrated upstream with increasing actuation strength, whereas the downstream stagnation point maintained a constant position. Bubble-type breakdown was stable under these conditions. The overall shape of the bubble did not change significantly with the strength of actuation increased from 18 kV to 21 kV. The interior structures of the bubble were nearly the same, containing one major cell or vortex ring. The gross exterior appearance of the bubble was axisymmetric, while the internal structure was asymmetric to some extent. The length-to-diameter ratio of the bubble was approximately 1.61 and 1.48, with the maximum diameter occurring at approximately Z = 20 mm and Z = 18 mm for 18 kV and 20 kV, respectively. With the increase of the excitation voltage, the asymmetry between the two different sides of the vortex ring increased, but the number of vortex rings did not increase.

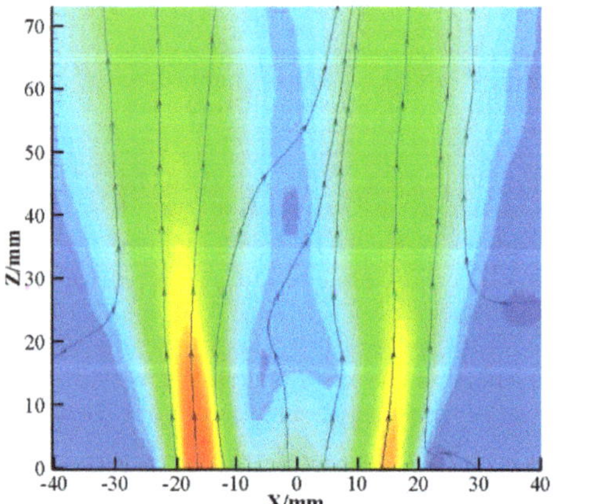

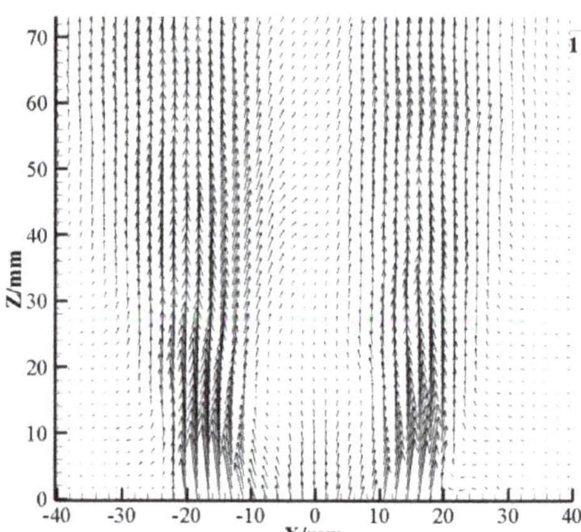

(**a**) DBD actuation off

Figure 3. *Cont.*

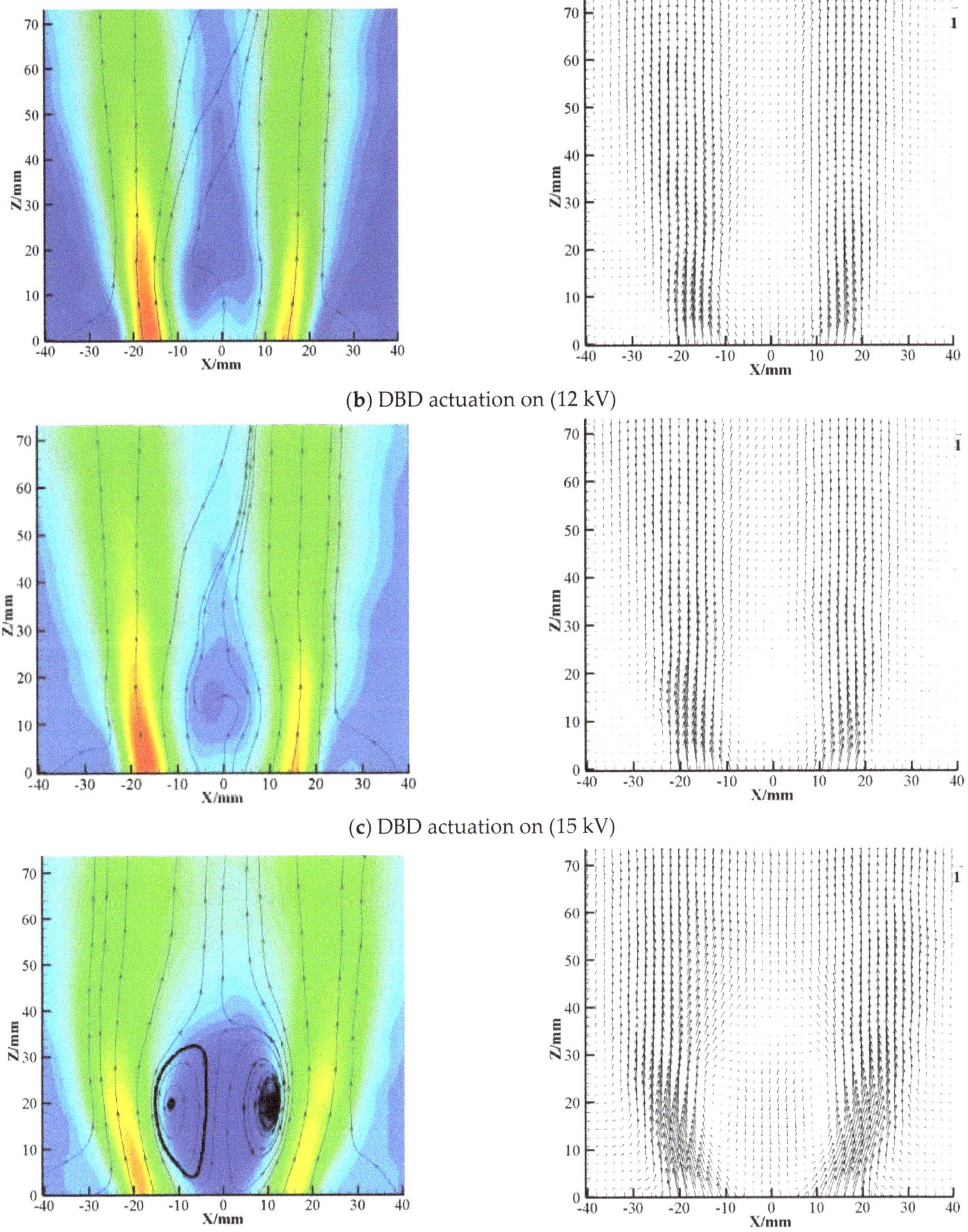

(**b**) DBD actuation on (12 kV)

(**c**) DBD actuation on (15 kV)

(**d**) DBD actuation on (18 kV)

Figure 3. *Cont.*

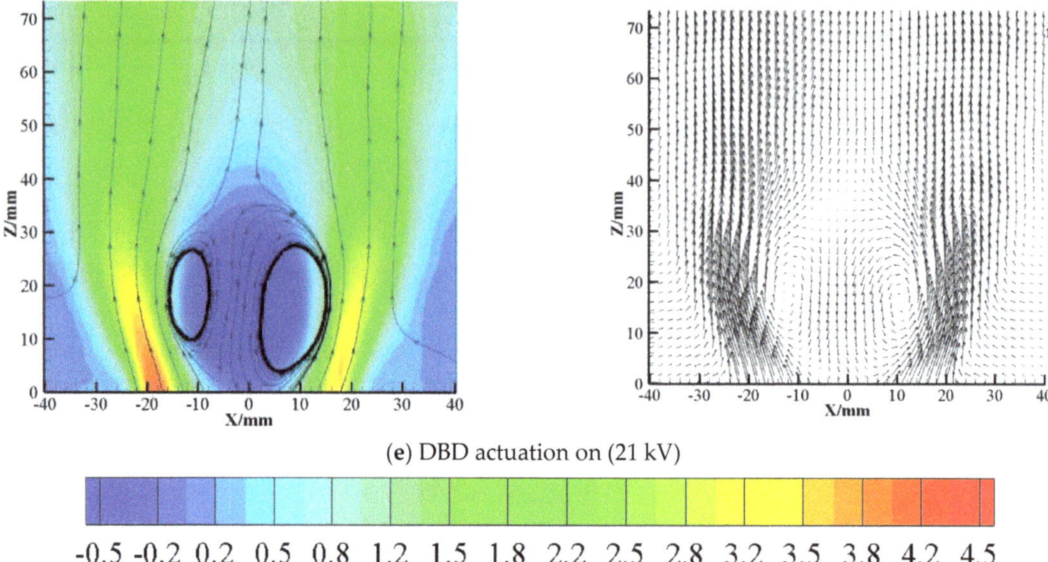

Figure 3. Contours of mean axial velocity magnitude, streamlines, and vectors (left: magnitude contours and streamlines; and right: vectors) in the XZ plane with bulk velocity of 3 m/s.

Figure 4 summarizes the axial velocity profiles along the centerline (X = 0 mm) for the five cases with voltages ranging from 0 kV to 21 kV. The figure shows that the gradual increase of voltage resulted in a gradual deceleration of the axial velocity along the centerline and a gradual appearance and upstream movement of the vortex breakdown. Without plasma actuation, the axial velocity at the injector outlet was 1.2 m/s. Moving downstream, the axial velocity linearly decreased to Z = 22 mm, achieving a lower value of 0.23 m/s. Then a velocity plateau appeared, which means the velocity remained essentially unchanged. Applying the 12 kV excitation, the axial velocity distribution in the Z = 0~20 mm zone moved down slightly, and the injector outlet velocity dropped to 0.99 m/s, but the overall change was not obvious, and the velocity plateau positions coincided. When the excitation was increased to 15 kV, the injector outlet velocity decreased to 0.62 m/s, and the axial velocity, in the range of Z = 0~20 mm, decreased appreciably. However, because the vortex breakdown zone was very small and did not pass through the jet centerline, the stagnation point and recirculation zone cannot be captured in the axial velocity distribution along the centerline. When the excitation was increased to 18 kV, the injector outlet velocity decreased to −0.1 m/s, and an obvious backflow zone appeared in the region of Z = 0~40 mm. When the excitation was further increased to 21 kV, the overall velocity distribution changed slightly. Although the magnitude of the reverse velocity increased slightly, the length of the vortex zone and the position of the stagnation point remained basically unchanged. The velocity increased linearly after it reached the minimum value at Z = 22 mm, and no velocity plateau could be observed.

Figure 5 shows the axial velocity magnitude contours and streamlines with a bulk velocity of 5.7 m/s. The plasma actuation-induced bubble-type vortex breakdown was still obvious, while the interior structures of the bubble contained one major cell as well. However, the size of the bubble decreased, which indicates that the effect of plasma actuation decreases with the increase of flow rate. More work is needed to establish the correlations between the effectiveness of plasma actuation, the jet inlet velocity, and other parameters, in order to gain a full understanding. Further research is needed to enhance the plasma flow control authority to put this method into engineering applications, including full-scale applications.

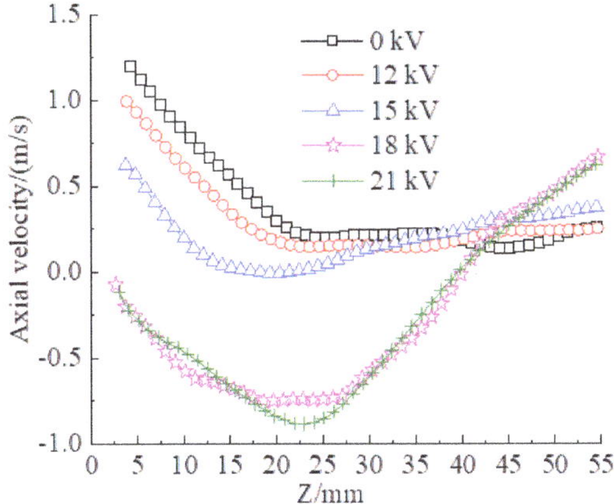

Figure 4. Profiles of the mean axial velocity.

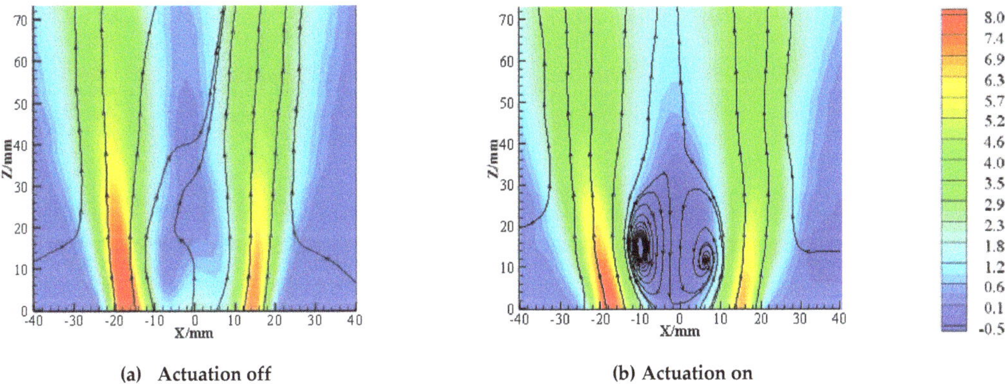

Figure 5. Contours of mean axial velocity magnitude and streamlinesin the XZ plane with bulk velocity of 5.7 m/s.

4. Conclusions

In this study, the concept of using the plasma swirler to control vortex breakdown was proposed and tested. It was observed in the experiments that, by varying the strength of the plasma actuation, the vortex breakdown region could be effectively controlled. The plasma swirler, as a nonintrusive, no-moving-part method of controlling vortex breakdown, was demonstrated experimentally. Flowfield measurement demonstrated that the plasma actuation was both effective and efficient in controlling the development of vortex breakdown. Experimental results showed that the 15 kV actuation was strong enough to trigger the onset of vortex breakdown. The overall shape of the bubble did not change significantly when the strength of actuation increased from 18 kV to 21 kV. The interior structures of the bubble were nearly the same, containing one major cell or vortex ring. Without involving moving parts or mass adding, the plasma actuation offers great flexibility in flow and combustion control. It offers a choice for fundamental research of vortex breakdown phenomena, where the actuation strength and frequency can be varied and controlled easily. Further research is needed to enhance the plasma flow control authority to put this method into engineering applications.

Author Contributions: Conceptualization, G.L.; validation, X.J., W.D. and J.Y.; formal analysis, C.L.; data curation, Y.M.; writing—original draft preparation, G.L.; writing—review and editing, X.J.; supervision, G.X. All authors have read and agreed to the published version of the manuscript.

Funding: This research received no external funding.

Acknowledgments: The research was supported by the National Natural Science Foundation of China (Grant No. 51676188) and the National Youth Natural Science Foundation of China (Grant No. 51706224), which are gratefully acknowledged.

Conflicts of Interest: The authors declare no conflict of interest.

References

1. Roth, J.R.; Sherman, D.M.; Wilkinson, P.S. Electrohydrodynamic flow control with a glow-discharge surface plasma. *AIAA J.* **2000**, *38*, 1166–1172. [CrossRef]
2. Moreau, E. Airflow control by non-thermal plasma actuators. *J. Phys. D Appl. Phys.* **2007**, *40*, 605–636. [CrossRef]
3. Corke, T.C.; Enloe, C.L.; Wilkinson, S.P. Dielectric Barrier Discharge Plasma Actuators for Flow Control. *Annu. Rev. Fluid Mech.* **2010**, *42*, 505–529. [CrossRef]
4. Wang, J.J.; Choi, K.S.; Feng, L.H.; Timothy, D.W. Richard, Recent developments in DBD plasma flow control. *Prog. Aerosp. Sci.* **2013**, *62*, 52–78. [CrossRef]
5. Kriegseis, J.; Simon, B.; Grundmann, S. Towards in-flight applications? A review on dielectric barrier discharge-based boundary-layer control. *Appl. Mech. Rev.* **2016**, *68*, 020802. [CrossRef]
6. Leonov, S.B.; Adamovich, I.V.; Soloviev, V.R. Dynamics of near-surface electric discharges and mechanisms of their interaction with the airflow. *Plasma Sources Sci. Technol.* **2016**, *25*, 063001. [CrossRef]
7. Konstantinidis, E. Active Control of Bluff-Body Flows Using Plasma Actuators. *Actuators* **2019**, *8*, 66. [CrossRef]
8. Roupassov, D.V.; Nikipelov, A.A.; Nudnova, M.M.; Starikovskii, A.Y. Flow Separation Control by Plasma Actuator with Nanosecond Pulsed-Periodic Discharge. *AIAA J.* **2009**, *47*, 168–185. [CrossRef]
9. Little, J.; Takashima, K.; Nishihara, M.; Adamovich, I.; Samimy, M. Separation Control with Nanosecond-Pulse-Driven Dielectric Barrier Discharge Plasma Actuators. *AIAA J.* **2012**, *50*, 350–365. [CrossRef]
10. Fujii, K. Three Flow Features behind the Flow Control Authority of DBD Plasma Actuator: Result of High-Fidelity Simulations and the Related Experiments. *Appl. Sci.* **2018**, *8*, 546. [CrossRef]
11. Shyy, W.; Jayaraman, B.; Andersson, A. Modeling of glow discharge-induced fluid dynamics. *J. Appl. Phys.* **2002**, *92*, 6434–6443. [CrossRef]
12. Hasan, M.; Atkinson, M. Investigation of a Dielectric Barrier Discharge Plasma Actuator to Control Turbulent Boundary Layer Separation. *Appl. Sci.* **2020**, *10*, 1911. [CrossRef]
13. Skourides, C.; Nyfantis, D.; Leyland, P.; Bosse, H.; Ott, P. Mechanisms of Control Authority by Nanosecond Pulsed Dielectric Barrier Discharge Actuators on Flow Separation. *Appl. Sci.* **2019**, *9*, 2989. [CrossRef]
14. Pescini, E.; de Giorgi, M.G.; Suma, A.; Francioso, L.; Ficarella, A. Separation control by a microfabricated SDBD plasma actuator for small engine turbine applications: Influence of the excitation waveform. *Aerosp. Sci. Technol.* **2018**, *76*, 442–454. [CrossRef]
15. Lo, K.-H.; Sriram, R.; Kontis, K. Wake flow characteristics over an articulated lorry model with/without AC-DBD plasma actuation. *Appl. Sci.* **2019**, *9*, 2426. [CrossRef]
16. Go, D.B.; Garimella, S.V.; Fisher, T.S.; Mongia, R.K. Ionic winds for locally enhanced cooling. *J. Appl. Phys.* **2007**, *102*, 053302. [CrossRef]
17. Roy, S.; Wang, C.-C. Plasma actuated heat transfer. *Appl. Phys. Lett.* **2008**, *92*, 231501. [CrossRef]
18. Audier, P.; Fenot, M.; Benard, N.; Moreau, E. Film cooling effectiveness enhancement using surface dielectric barrier discharge plasma actuator. *Int. J. Heat Fluid Flow* **2016**, *62*, 247–257. [CrossRef]
19. Xiao, Y.; Dai, S.; He, L.; Jin, T.; Zhang, Q.; Hou, P. Investigation of film cooling from cylindrical hole with plasma actuator on flat plate. *Heat Mass Transf.* **2016**, *52*, 1571–1583. [CrossRef]
20. Kim, Y.J.; Kim, G.M.; Shin, Y.; Kwak, J.S. Experimental Investigation on the Effects of DBD Plasma on the Film Cooling Effectiveness of a 30-Degree Slot. *Appl. Sci.* **2017**, *7*, 633. [CrossRef]
21. Uehara, S.; Takana, H. Surface cooling by dielectric barrier discharge plasma actuator in confinement channel. *J. Electrost.* **2020**, *104*, 103417. [CrossRef]
22. Hebrero, F.C.; Adamo, J.D.; Sosa, R.; Artana, G. Vortex induced vibrations suppression for a cylinder with plasma actuators. *J. Sound Vib.* **2020**, *468*, 115121. [CrossRef]
23. Yokoyama, H.; Tanimoto, I.; Iida, A. Experimental Tests and Aeroacoustic Simulations of the Control of Cavity Tone by Plasma Actuators. *Appl. Sci.* **2017**, *7*, 790. [CrossRef]
24. de Jong, A.; Bijl, H. Corner-type plasma actuators for cavity flow-induced noise control. *AIAA J.* **2014**, *52*, 33–42. [CrossRef]
25. da Silva, G.P.G.; Eguea, J.P.; Croce, J.A.G.; Catalano, M.F. Slat aerodynamic noise reduction using dielectric barrier discharge plasma actuators. *Aerosp. Sci. Technol.* **2020**, *97*, 105642. [CrossRef]
26. Sato, S.; Yokoyama, H.; Iida, A. Control of Flow around an Oscillating Plate for Lift Enhancement by Plasma Actuators. *Appl. Sci.* **2019**, *9*, 776. [CrossRef]
27. Motta, V.; Malzacher, L.; Peitsch, D. Numerical Assessment of Virtual Control Surfaces for Load Alleviation on Compressor Blades. *Appl. Sci.* **2018**, *8*, 125. [CrossRef]

28. Roy, S.; Wang, C.-C. Bulk flow modification with horseshoe and serpentine plasma actuators. *J. Phys. D Appl. Phys.* **2008**, *42*, 032004. [CrossRef]
29. Hoskinson, A.R.; Hershkowitz, N.; Ashpis, D.E. Force measurements of single and double barrier DBD plasma actuators in quiescent air. *J. Phys. D Appl. Phys.* **2008**, *41*, 245209. [CrossRef]
30. Pescini, E.; Francioso, L.; De Giorgi, M.G.; Ficarella, A. Investigation of a micro dielectric barrier discharge plasma actuator for regional aircraft active flow control. *IEEE Trans. Plasma Sci.* **2015**, *43*, 3668–3680. [CrossRef]
31. Moreau, E.; Cazour, J.; Benard, N. Influence of the air-exposed active electrode shape on the electrical, optical and mechanical characteristics of a surface dielectric barrier discharge plasma actuator. *J. Electrost.* **2018**, *93*, 146–153. [CrossRef]
32. Benard, N.; Audier, P.; Moreau, E.; Takashima, K.; Mizuno, A. Active plasma grid for on-demand airflow mixing increase. *J. Electrost.* **2017**, *88*, 15–23. [CrossRef]
33. Peckham, D.H.; Atkinson, S.A. *Preliminary Results of Low Speed Wind Tunnel Test on a Ghotic Wing of Aspect Ratio 1.0*; British Aeronautical Research Council: London, UK, 1957.
34. Sarpkaya, T. On stationary and travelling vortex breakdowns. *J. Fluid Mech.* **1971**, *45*, 545–559. [CrossRef]
35. Sarpkaya, T. Vortex Breakdown in Swirling Conical Flows. *AIAA J.* **1971**, *9*, 1792–1799. [CrossRef]
36. Leibovich, S. The Structure of Vortex Breakdown. *Annu. Rev. Fluid Mech.* **1978**, *10*, 221–246. [CrossRef]
37. Benjamin, T.B. Theory of the vortex breakdown phenomenon. *J. Fluid Mech.* **1962**, *14*, 593–629. [CrossRef]
38. Gartshore, I.S. *Recent Work in Swirling Incompressible Flow*; Report LR-343; National Research Council Canada: Ottawa, OT, Canada, 1962.
39. Leibovich, S.; Stewartson, K. A sufficient condition for the instability of columnar vortices. *J. Fluid Mech.* **1983**, *126*, 335–356. [CrossRef]
40. Shtern, V. *Cellular Flows*; Cambridge University Press: New York, NY, USA, 2018.
41. Althaus, W.; Krause, E.; Hofhaus, J.; Weimer, M. Vortex breakdown: Transition between bubble- and spiral-type breakdown. *Meccanica* **1994**, *29*, 373–382. [CrossRef]
42. Herrada, M.A.; Shtern, V. Control of vortex breakdown by temperature gradients. *Phys. Fluids* **2003**, *15*, 3468–3477. [CrossRef]
43. Srigrarom, S.; Kurosaka, M. Shaping of delta-wing planform to suppress vortex breakdown. *AIAA J.* **2012**, *38*, 183–186. [CrossRef]
44. Schmucker, A.; Gersten, K. Vortex breakdown and its control on delta wings. *Fluid Dyn. Res.* **1998**, *3*, 268. [CrossRef]
45. Gutmark, E.J.; Guillot, S.A. Control of vortex breakdown over highly swept wings. *AIAA J.* **2005**, *43*, 2065–2069. [CrossRef]
46. Mununga, L.; Jacono, D.L.; Sorensen, J.N.; Leweke, T.; Thompson, M.C.; Hourigan, K. Control of confined vortex breakdown with partial rotating lids. *J. Fluid Mech.* **2014**, *738*, 5–33. [CrossRef]
47. Husain, H.S.; Shtern, V.; Hussain, F. Control of vortex breakdown by addition of near-axis swirl. *Phys. Fluids* **2003**, *15*, 271–279. [CrossRef]
48. Zheltovodov, A.; Pimonov, E.; Knight, D. Supersonic Vortex Breakdown Control by Energy Deposition. In Proceedings of the 43rd AIAA Aerospace Sciences Meeting and Exhibit, Reno, NV, USA, 10–13 January 2005; p. 1048.
49. Mitchell, A.M.; Delery, J. Research into vortex breakdown control. *Prog. Aerosp. Sci.* **2001**, *37*, 385–418. [CrossRef]
50. Li, G.; Jiang, X. Effects of electrical parameters on the performance of a plasma swirler. *Phys. Scr.* **2019**, *94*, 095601. [CrossRef]
51. Li, C.; Shao, W.; Xu, Y.; Hu, H.; Liu, Y.; Nie, C.; Zhu, J. Swirl diffusion flame control by the plasma swirler. *Sci. China Ser. E Technol. Sci.* **2011**, *54*, 1820–1825. [CrossRef]
52. Li, G.; Jiang, X.; Zhao, Y.; Liu, C.; Chen, Q.; Xu, G.; Liu, F. Jet flow and premixed jet flame control by plasma swirler. *Phys. Lett. A* **2017**, *381*, 1158–1162. [CrossRef]
53. Li, G.; Jiang, X.; Zhu, J.; Yang, J.; Liu, C.; Mu, Y.; Xu, G. Combustion control using a lobed swirl injector and a plasma swirler. *Appl. Therm. Eng.* **2019**, *152*, 92–102. [CrossRef]
54. Li, G.; Jiang, X.; Chen, Q.; Wang, Z. Flame lift-off height control by a combined vane-plasma swirler. *J. Phys. D Appl. Phys.* **2018**, *51*, 345205. [CrossRef]
55. Li, G.; Jiang, X.; Jiang, L.; Lei, Z.; Zhu, J.; Mu, Y.; Xu, G. Design and experimental evaluation of a plasma swirler with helical shaped actuators. *Sens. Actuators A Phys.* **2020**, *315*, 112250. [CrossRef]

Article

Influence of Ag Electrodes Asymmetry Arrangement on Their Erosion Wear and Nanoparticle Synthesis in Spark Discharge

Kirill Khabarov *, Maxim Urazov, Anna Lizunova, Ekaterina Kameneva, Alexey Efimov and Victor Ivanov

Department of Physical and Quantum Electronics, Moscow Institute of Physics and Technology, National Research University, 141701 Dolgoprudny, Russia; desperate@inbox.ru (M.U.); Lizunova.aa@mipt.ru (A.L.); katerinakamenev@yandex.ru (E.K.); efimov.aa@mipt.ru (A.E.); ivanov.vv@mipt.ru (V.I.)

* Correspondence: kirill.khabarov@phystech.edu

Abstract: For nanoparticle synthesis in a spark discharge, the influence of the degree of electrode asymmetry in the rod-to-rod configuration, using the example of silver electrodes, on the energy efficiency and nanoparticle composition is studied. The asymmetry degree was determined by the angle between electrodes' end faces. Two types of discharge current pulses were used: oscillation-damped and unipolar, in which electrodes changed their polarities and had a constant polarity during a single discharge, respectively. A significant influence of the asymmetry degree of the electrode arrangement on the synthesized nanoparticle size, agglomeration and concentration, and on the synthesis energy efficiency, has been established. An increase in the degree of the electrode asymmetry with the oscillation-damped discharge current pulse led to an increased mass production rate and energy efficiency of nanoparticle synthesis, a significant fraction of which had large dimensions of more than 40 nm. The effect of the transfer of synthesized nanoparticles to the opposite electrode at the unipolar discharge current pulse led to the appearance of electroerosive instability, manifested in the formation of a protrusion on the anode surface, around which spark discharges, leading to its further growth and electrode gap closure.

Keywords: spark discharge; nanoparticle synthesis; silver electrodes; electrodes asymmetry

Citation: Khabarov, K.; Urazov, M.; Lizunova, A.; Kameneva, E.; Efimov, A.; Ivanov, V. Influence of Ag Electrodes Asymmetry Arrangement on Their Erosion Wear and Nanoparticle Synthesis in Spark Discharge. *Appl. Sci.* **2021**, *11*, 4147. https://doi.org/10.3390/app11094147

Academic Editor: Mariusz Jasiński

Received: 29 March 2021
Accepted: 29 April 2021
Published: 1 May 2021

Publisher's Note: MDPI stays neutral with regard to jurisdictional claims in published maps and institutional affiliations.

1. Introduction

A spark discharge (SD) [1,2] is a good method for producing nanoparticles (NPs) of small size (less than 20 nm) by the electrical erosion [1] of the electrode material [3] with a fast response due to instantaneous changes in the parameters of the electrical discharge circuit [4]. In the SD processes, high-energy spark discharges are created between two electrodes in a controlled gasflow, eroding the electrode material and forming NPs in the form of aerosol [5]. A large set of SD parameters makes it possible to fine-tune the shape, size, and concentration of the synthesized NPs. For instance, this method can be applied for NP synthesis for the purpose of formation and doping of multicomponent structures: metal alloys [6,7], semiconductor quantum dots [8–11], optical and magnetic materials [12], and high-temperature superconductors [13,14]. Additionally, the method is convenient for manufacturing devices for microelectronics and photonics [15], gas [16,17] and biological sensors [18] based on NPs, microconductors, resistors [19,20] and capacitive elements [2]. However, incorrect selection of method parameters leads to non-optimal use of the discharge energy [21] and the production of large parasitic particles (more than 40 nm).

SD parameters, determining the diversity of NPs and the stability of their synthesis, have been studied by different methods, but there are still questions that require detailed consideration. Thus, the authors of [4,22] studied the size and shape dependence of the synthesized particles from the discharge pulse energy, frequency of discharge repetition, and parameters of the carrier gas, such as the composition, flow rate, pressure, and temperature of the gas medium [23–25]. For example, the energy released in the discharge and the pulse repetition rate are both key parameters that strongly influence the size, shape, and concentration of the synthesized NPs [8]. However, determining the exact value of the energy in the discharge is a complex experimental task, solved by measuring the current and voltage in the interelectrode gap in SD processes for NP synthesis [4] and for micro-electro discharge machining (EDM) [26,27]. The authors of [4] were measuring pulsed voltage across the interelectrode gap and proposed an algorithm for determining the energy transferred into the discharge gap. In the paper [26] Yang et al. received the spark discharge waveforms in the gap and presented a model for the micro-EDM parameters' prediction. D'Urso et al. [27] performed a method for the evaluation of discharge parameters and investigated their influence and the influence of electrode size on stainless steel drilling by micro-EDM. A simpler approach is to estimate the energy released in the spark gap by approximating the active impedance of the gap with an equivalent constant electrical resistance [1,5].

Additionally to the energy parameters, the influence of the electrodes' relative position and the gas flow blowing past them on the synthesis of particles is of great interest. There are many configurations of the electrode arrangement for spark discharge generators (SDG): rod-to-rod, rod-to-tube, pin-to-plate, wire-in-hole, and wire-to-hole [28,29]. Each of them requires precise adjustment and control of the electrodes' positions. For instance, in wire-in-hole and wire-to-hole configurations with a wire thickness of about 0.5 mm, it is quite difficult to stabilize the position of the wire electrode in the gas flow. Thus, in the work [30], electrodes were constantly adjusted to prevent the electrical circuit break. Currently, the rod-to-rod configuration is the most common. It involves a pair of cylindrical electrodes fixed in crimping holders and separated by a gap [31]. At the same time, this configuration is the least demanding for the electrodes' setup accuracy for the purpose of particle synthesis repeatability. The influence of the electrodes' position relative to the gas flow on the size and shape of particles and their agglomerates was studied experimentally and by computer modeling methods by the authors of [32]. It was also found that the diameter of rod-shaped electrodes significantly affects the energy efficiency of NP synthesis [21].

In this paper, we study the influence of another important parameter on the processes of NP synthesis in the SD—the degree of asymmetry of the electrode arrangement in the rod-to-rod configuration on the example of silver electrodes. The two types of discharge current pulses were studied—oscillation-damped, in which the electrodes changed their polarities during a single discharge, and unipolar, in which the electrodes had a given polarity during the discharge. Electrodes in the form of rods, one of which had a gas supply hole, were installed coaxially. The end faces of the installed electrodes, from which the electrical erosion of the material was performed during the discharge current flow, were oriented in parallel and at an angle. Here, the parameters of the gas flow and the capacitor initial energy were the same in these experiments.

2. Materials and Methods

In this paper, we study the changes in the size and morphology of silver NPs obtained in the SD on different setup variants for electrodes' working surfaces and on changes in parameters of the SDG electrical circuit. A schematic representation of the SDG setup is shown in Figure 1a. The SD was formed between two coaxially exposed cylindrical silver electrodes with a diameter of 8 mm, along the axis of one of which a cylindrical hole with a diameter of 1.4 mm was made to supply a gas flow into the interelectrode gap. The distance between the two nearest points of electrodes was initially set to 1 mm. The electrical part of the SDG circuit was implemented in two different versions—with an additional ballast resistor with a nominal resistance of 5.0 Ω, causing a unipolar discharge current pulse (UDP), and without it, causing an oscillation-damped discharge current pulse (ODP). In each case, the voltage waveform on the capacitor during a discharge was recorded by the oscilloscope DPO4102B-L (Tektronix, Beaverton, OR, USA) with an upper limit frequency of 20 MHz on a fast scan from a RC-voltage divider with an upper limit frequency of about 7 MHz. Since the ODP and UDP correspond to the oscillations with a positive value of the square of the present frequency $\omega^2 = \omega_0^2 - \gamma^2 > 0$ (with $\omega^2 \to +0$ for the UDP case), the signal form is well described by the damped cosine function in the experiment (Figure 2) [33]:

$$U = U_c e^{-\gamma t} \cos \omega t. \quad (1)$$

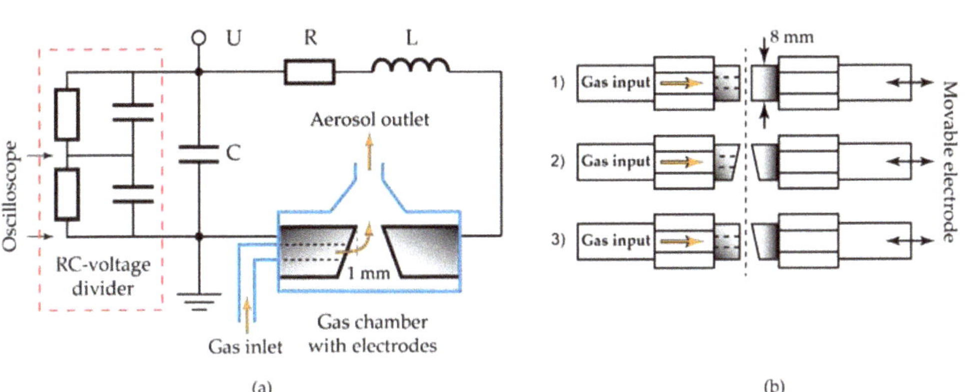

Figure 1. Equivalent representation of (**a**) the SDG setup and (**b**) of electrodes with the different position of their ends: (**1**) installed in parallel and at an angle for the experiments with (**2**) the ODP and (**3**) the UDP.

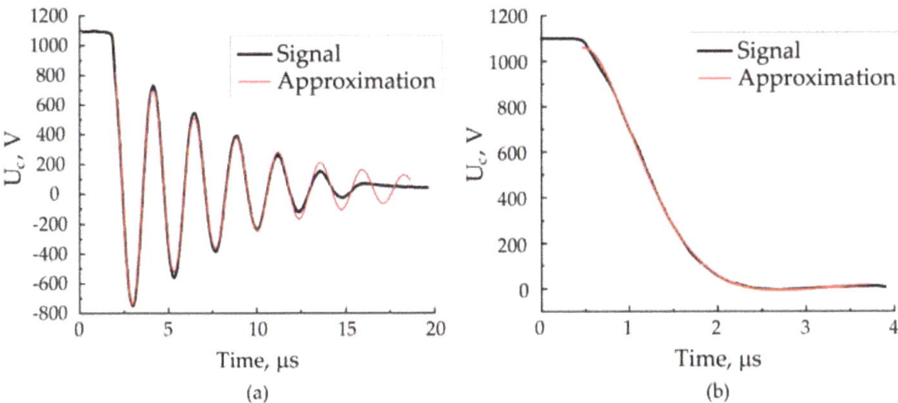

Figure 2. Waveforms of the voltage pulse (**a**) without and (**b**) with a ballast resistor in the electrical circuit for the symmetrical position of electrodes.

Here and next, U_c stands for the voltage amplitude, t for time, $\gamma = R_e(2L)^{-1}$ for the damping coefficient, $\omega_0 = (LC)^{-0.5}$ for the natural frequency of harmonic oscillations, $R_e = R + R_s$ for the total equivalent electrical resistance of the discharge circuit, represented as the sum of the equivalent electrical resistances of the discharge circuit R and of the spark gap R_s, and L and C for circuit inductance and capacitance, respectively.

NP synthesis during the erosion of electrodes was done in a T-shaped chamber made of refractory glass (Duran, Milville, NJ, USA) in an argon atmosphere (99.9999%) at an excess pressure of 10 kPa, recorded by a pressure gauge, with a gas flow rate of 200 mL/min, recorded by the flow meter RRG-12 (Eltochpribor, Zelenograd, Russia).

The experiments were conducted in the mode of the interelectrode gap electrical breakdown. The nominal value of the discharge capacitor was C = 107 nF, the initial values of the capacitor charging voltage and frequency of capacitor discharges were U_c = 1100 V and ν = 500 Hz, respectively. In the case of the ODP, the electrical resistance R and the inductance L of the discharge circuit, without active loss in the interelectrode gap, were determined by the corresponding parasitic values in wires and in the capacitor. In the case of the UDP, a ballast-cooled resistor with a nominal value of 5.0 Ω was additionally installed. Approximating the discharge voltage waveforms for the symmetrical position of electrodes (Figure 2) with the function (1), the unknown equivalent total electrical resistance and inductance of the discharge circuit, including the spark gap, were L = 1.31 μH, $R_e = 0.40$ Ω and L = 1.97 μH, $R_e = 6.16$ Ω for the ODP and UDP, respectively. Electrical resistances of the discharge circuit R = 0.13 Ω and 5.55 Ω with the ODP and UDP, respectively, were measured at the natural frequency of discharge current oscillations by the impedance spectrometer RLC-meter AMM-3058 (AKTAKOM, Santa Clara, CA, USA). In this case, current and potential contacts of the device were connected to symmetrically arranged electrodes at ends facing each other. Taking into account these measurements, equivalent electrical resistances of the interelectrode spark gap were 0.27 Ω and 0.61 Ω for the ODP and UDP, respectively.

At the beginning of each experiment, electrodes' surfaces, intended for the electroerosive NP synthesis, were polished on the grinding and polishing machine Digiprep Accura (Metkon, Bursa, Turkey) to a roughness of Ra = 0.5 μm. The average electrode roughness was measured by the optical profilometer S neox (Sensofar, Terrassa, Spain) on the confocal 20× lens from the area 800 × 800 μm^2.

In the experiments, electrodes were arranged coaxially and their nearest end faces were located parallel (at the angle of 0°) and at the angles of 7° and 15° for the ODP and parallel (at the angle of 0°) and at the angles of 3° and 10° for the UDP. Assuming that the electrode with a positive potential does not undergo significant erosion in the case of the UDP, only cathode end face was beveled, whereas in the case of the ODP, ends of both electrodes were set at an angle (Figure 1b).

NPs were synthesized in the process of electrode erosion by a pulse-periodic SD continuously for 6 h maximum. During this experiment, we studied the statistical size distribution of silver NP agglomerates in the aerosol stream by their differential electrical mobility via the aerosol NP analyzer SMPS 3936 (TSI Inc., Shoreview, MN, USA). The statistical agglomerate size distribution was described by a log-normal distribution. The time dependence of the NP agglomerates modal size during electrode wear was constructed from distributions maxima. In parallel, we recorded changes in the capacitor charging voltage and the frequency of capacitor discharges in the setup by oscillograph readings. In addition, samples of the synthesized NPs were studied by the transmission electron microscope (TEM) JEM-2100 (JEOL, Ltd., Tokyo, Japan) for each regime under the study. For that purpose they were deposited on TEM grids in 5–10 min intervals from the start of experiments and after the establishment of thermodynamic equilibrium in the spark gap. Based on the TEM images, we collected statistics on the NPs size for at least 1000 pieces approximating them by round particles of equivalent cross-sectional area and drew their size distribution. At the end of experiments, we investigated electrodes' end faces after their electrical erosion, and their elemental composition in the scanning electron microscope (SEM) JSM 7001F (JEOL, Ltd., Tokyo, Japan) with the option of energy dispersive X-ray spectroscopy (EDX).

At the beginning and in the end of each experiment, we weighed electrodes on the scale SQP-F SECURA 225D-1ORU (Sartorius Lab Instruments, Goettingen, Germany) with a resolution of 10 μg to estimate the electrodes' mass loss as a result of their erosion.

3. Results

Installing electrodes in the synthesis chamber of SDG, it is difficult to avoid some asymmetry in their relative positions. The appearance of random angles between the planes of erosion ends of electrodes leads to a change in the composition of synthesized NPs and to non-reproducible results. In the present study, we purposely set a certain degree of electrode asymmetry and evaluated its effect on the size, shape and concentration of NPs, as well as the energy efficiency of synthesis with the ODP in which the electrodes changed their polarities during a single discharge, and UDP, in which the electrodes had a given polarity during the discharge.

3.1. The Influence of the Degree of Electrodes Asymmetry on the NP Synthesis with the ODP

The characteristic side view of an electrode pair installed in the discharge chamber, with end faces located at the angle of 15° to each other, is shown in Figure 3 for the three cases of the ODP: the initial position before the SD process, during the SD process, and 6 h after the SD process. As it can be seen from Figure 3b, the SD is localized in the region of the shortest distance between the electrodes, which is characteristic of the electric breakdown of gas gaps. Figure 3c shows a significant wear of the beveled end faces of both electrodes during 6 h of processing: the ends were aligned during this time to a parallel state. Photos of the electrodes end faces, SEM images of their surface characteristic areas (types 1–3 in Figure 4a–c) and chemical composition of the obtained surfaces for the cases of electrodes parallel location and location at the angle of 15° are shown in Figure 4. Moreover, the surface characteristic areas of type 3 are similar to each other for the cases of parallel orientation (Figure 4b) and at an angle (Figure 4c).

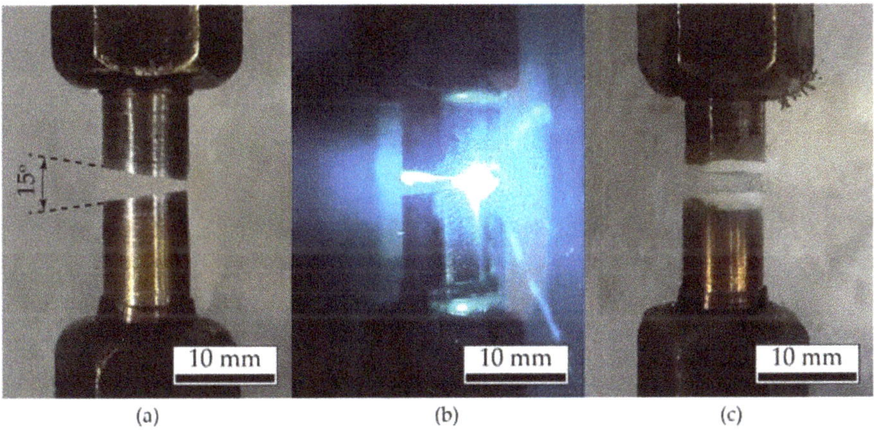

Figure 3. Photos of the gap between the electrodes with end faces forming the angle of 15°: (**a**) before the start of the SD process, (**b**) during the discharge process and (**c**) 6 h after the spark treatment by the ODP.

The SEM image of area 1 of Figure 4a shows a typical surface of the original polished electrode, characterized by an average roughness of Ra = 0.5 μm. On the electrodes' end faces, initially located at the angle of 15° (Figure 4b), after processing with the SD, two typical areas marked in the Figure 4b with the numbers 2 and 3 are observed. As it can be seen from the SEM image of the area 2, the polished surface of the electrode is covered with a loose layer of dendrite-like NP agglomerates. Presumably, the main part of such agglomerates was formed with electrical erosion and transferred from the opposite electrode. Area 3 represents exactly the electrode surface treated with the SD, and as it can be seen from the SEM image, it consists of overlapping craters with dimensions of about 10 μm. Each subsequent discharge, leading to the formation of a new crater, can occur in places of increased electric field density on microprotrusions formed over the previous crater edge as a result of spilling molten metal. The elemental composition of Ag electrode surfaces, according to the results of EDX (Figure 4g–i), dominantly consist of silver for all cases, presented in Figure 4a–c. In this case, the presence of carbon may be associated with the presence of carbon-containing compounds in the carrier gas and adsorbed substances in the gas discharge chamber and on the initial electrodes.

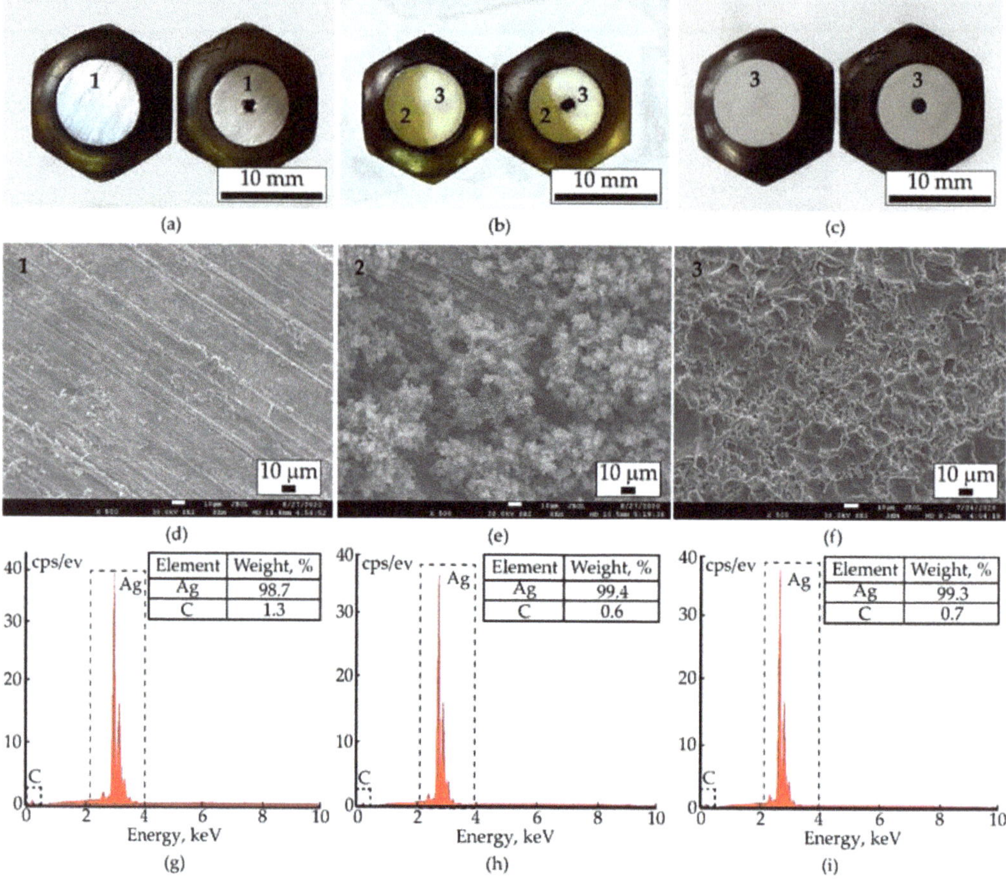

Figure 4. (**a**) Photos of the polished electrodes, (**b**) the electrodes after 6 h of NP synthesis with the ends initially located at the angle of 15° and (**c**) with the initially parallel ends, (**d–f**) SEM images of the corresponding characteristic areas 1, 2, 3, indicated in the photos of the electrode ends, and (**g–i**) EDX spectra of Ag electrode surfaces marked 1–3 in (**a–c**).

For the three cases of the mutual arrangement of electrodes end faces (in parallel and at the angles of 7° and 15°), time dependences of modal distribution sizes of NP agglomerates and their concentrations, determined by electrical mobility during 6 h of NP synthesis, were studied (Figure 5).

A characteristic of the measured dependences of agglomerates' modal dimensions from time is their asymptotic approximation to the average size of about 150 nm during 6 h of NP synthesis for each of the three cases of mutual arrangement of electrodes' end faces. Such an asymptotic behavior of the average agglomerate size–time–function indicates the stability of the SD process on asymmetrically installed electrodes, which leads to the formation of parallel end faces. The dependences comparison for the three cases shows that the average initial agglomerates' size and the average curvature of the function are higher with a greater degree of the electrodes' asymmetry. These features can be explained by a higher wear rate of electrodes with the end faces located at a large angle, due to their electrical erosion.

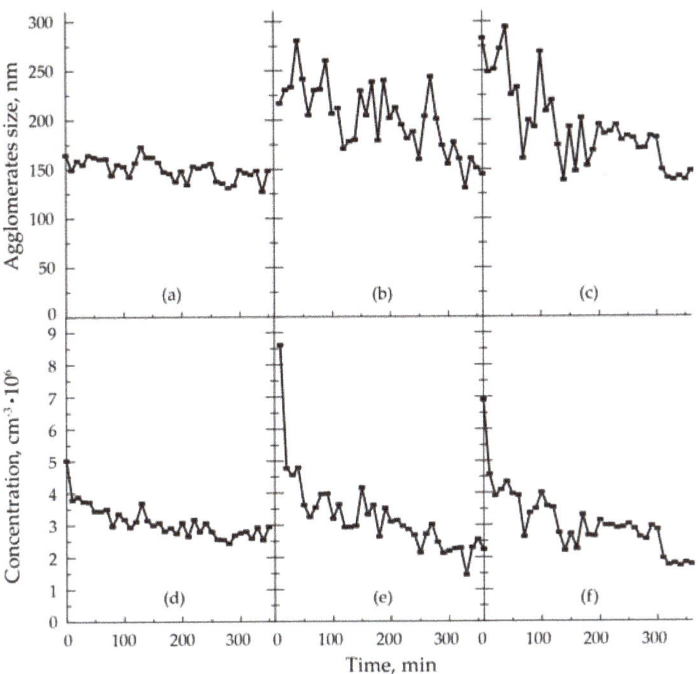

Figure 5. Time dependences of the modal size of NP agglomerates and their concentrations during the electrodes wear by the ODP for electrodes with ends (**a**,**d**) located parallel and at the angles of (**b**,**e**) 7° and (**c**,**f**) 15°.

The behavior of the dependences of the capacitor charging voltage and frequency of capacitor discharges with the ODP for the three cases of the electrodes' end arrangement, as it can be seen from Figure 6, has a common feature after the start of charge-discharge processes. Within first 5 min, there is an almost-identical decrease in the capacitor charging voltage and a simultaneous increase in the discharge frequency by about 35%. This effect is logically associated with the establishment of a stationary high-temperature regime in the spark gap and in electrodes. Next, the charging voltage curves monotonically increase and the discharge repetition rate curves monotonically decrease faster as the initial angle between electrodes' end faces increases. This effect can be associated with the erosion wear of the beveled electrodes end faces and, consequently, with a more rapid increase in the interelectrode gap distance in comparison with the case of parallel setup of the electrodes' ends. This behavior correlates well with increased modal size of NP agglomerates and their concentrations with an increase in the angle between electrodes' ends (Figure 5). We also note that the configuration of the electrodes' arrangement with the parallel ends gives the most stable behavior of the dependences of the capacitor charging voltage, frequency of discharges, and modal size of synthesized NP agglomerates and their concentrations.

It is also of interest to compare characteristic dimensions of craters formed by the SD on surfaces of electrodes initially installed parallel and at the angle 15° (Figure 7a,b). SEM images of the electrode surfaces being under erosion for 2 h were analyzed. The analysis was carried out only for the upper craters with a closed outer border in the form of a breastwork. As it can be seen from the comparison of the crater size distributions (Figure 7c,d), in the case of the initial parallel arrangement of electrodes end faces, the size of craters is noticeably smaller than when the electrodes end faces are located at an angle. This effect is logically explained by a smaller equivalent working area of electrodes' initially installed at an angle, and, as a result, a higher average discharge current density. At the same time, the craters' average diameter is less than 20 μm in both cases, corresponding to experimental observations in previously performed works of other authors [34–36].

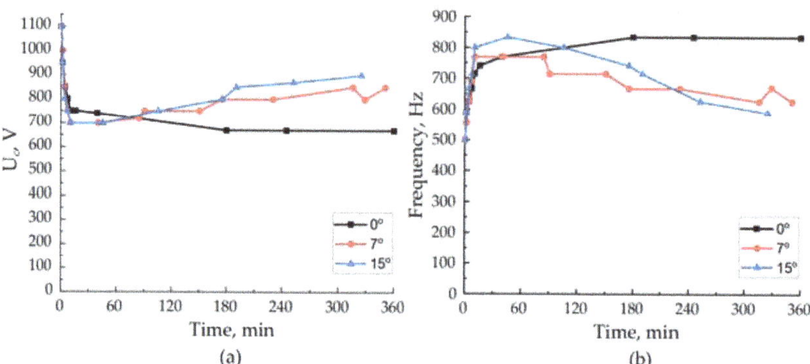

Figure 6. The dependences of (**a**) the capacitor charging voltage and (**b**) frequency of the capacitor discharges for the case of synthesis with the ODP for electrodes' ends located parallel and at angles of 7° and 15°.

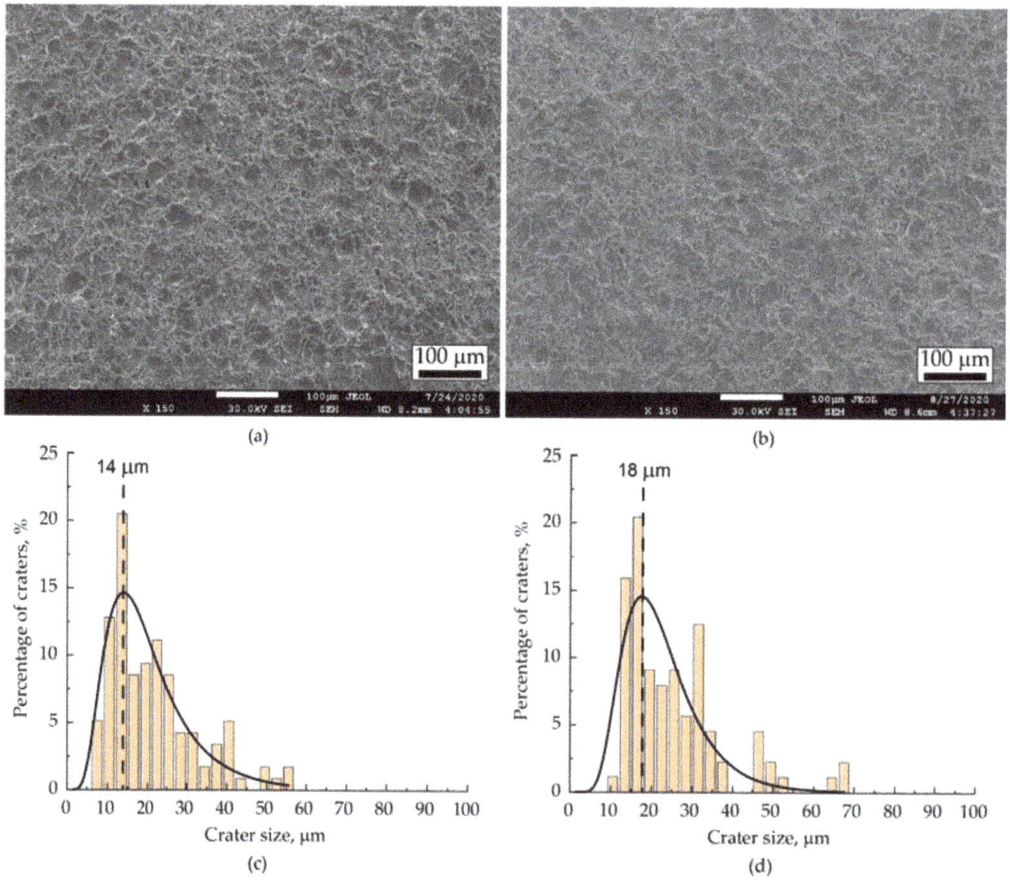

Figure 7. (**a**,**b**) SEM images of electrode faces subjected to electrical erosion and (**c**,**d**) the size distributions of craters for the cases of (**a**,**c**) the initial parallel arrangement of ends and (**b**,**d**) at the angle of 15°.

The sizes of synthesized primary NPs collected on TEM grids were analyzed on a series of TEM images (Figure 8a–c) for the three cases of the initial arrangement of electrodes end faces: parallel and at the angles of 7° and 15°. For this purpose, we obtained number-size distributions of NPs as part of agglomerates, and next converted them into mass-size distributions (Figure 8e,f), which clearly reflect the mass fraction of particles of different sizes.

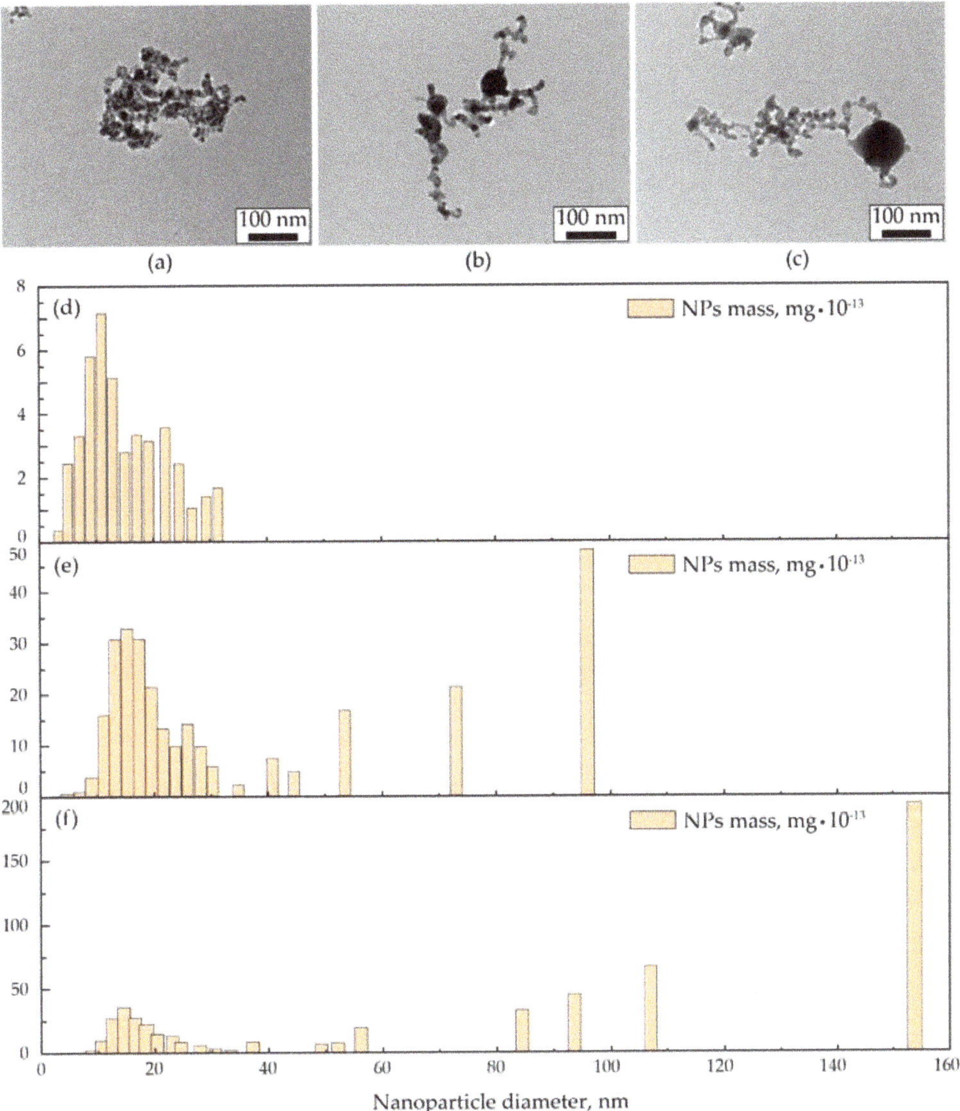

Figure 8. (**a–c**) TEM images and (**d–f**) mass distributions of NPs for the synthesis with the ODP and electrodes ends arranged (**a**,**d**) parallel and at the angles of (**b**,**e**) 7° and (**c**,**f**) 15°.

The given mass distributions qualitatively and quantitatively characterize the size and shape of synthesized NPs and their agglomerates for the three variants of the electrode ends arrangement. When going from the case of the parallel end faces arrangement to the case of their setup at a large angle, the NPs agglomeration increases, individual NPs form necks and NP agglomerates transform into aggregates. Additionally, despite the increase in the proportion of particles with a diameter up to 40 nm, the number of large parasitic particles with a size larger than 40 nm increases with an increase in the asymmetry degree of the electrodes' arrangement (Figure 8f). The mass fraction of parasitic particles turns out to be significant and equals 33% and 66% for the location of electrodes' ends at the angles of 7° and 15°, respectively.

When calculating the energy efficiency of the SDG, it was taken into account that the capacitor charging voltage $U_c(t)$ and the discharge frequency $\nu(t)$ are the values changing over the time of NP synthesis (Figure 7). Therefore, the total energy periodically stored in the capacitor and transferred to the discharge circuit during the NP synthesis operation time t was determined by the next formula [33]:

$$E_t = \int_0^t \frac{CU_c^2(t)}{2}\nu(t)dt. \quad (2)$$

To determine the part of the energy released directly in the spark gap, one can use the equivalent representation of the spark gap by the constant electrical resistance R_s. In this case, the total energy released per single discharge pulse of length τ can be represented as a sum of two terms: the energy release in elements of the discharge circuit and the energy release in the spark gap:

$$E_{pulse} = R\int_0^\tau I^2 dt + R_s \int_0^\tau I^2 dt. \quad (3)$$

A similar approach of calculating the discharge energy parameters was previously used in the works [1,26]. Using Formula (3), total energy released in the interelectrode gap during the whole SD process is calculated as follows:

$$E_t^s = E_t \frac{R_s}{R_e}. \quad (4)$$

The energy efficiency characterizing the useful output of the SDG is the synthesized NPs mass produced per unit useful energy consumed and expressed by the formula [1]:

$$m_e^d = \frac{\Delta M}{E_t^s}, \quad (5)$$

where $\Delta M = \Delta M_1 + \Delta M_2$ is the total mass of the synthesized NPs, equal to the mass loss of the two electrodes as a result of the NPs synthesis for time t. The total NPs mass after 6 h of synthesis, and the corresponding energy efficiency for the ODP at different arrangements of electrodes' end faces are presented in Table 1.

Table 1. The decrease in the electrodes' mass, the energy release in the interelectrode gap, and the energy efficiency of synthesis with the ODP.

Angle °	ΔM_1, mg	ΔM_2, mg	ΔM, mg	E_t^s, kJ	m_e^s, $\frac{\mu g}{kJ}$
0	12.90 ± 0.01	21.87 ± 0.01	34.77 ± 0.01	304 ± 3	114 ± 1
7	24.29 ± 0.01	47.34 ± 0.01	71.63 ± 0.01	316 ± 3	226 ± 2
15	30.12 ± 0.01	47.28 ± 0.01	77.40 ± 0.01	323 ± 3	239 ± 2

The maximum mass consumption of electrodes was observed for the ends installed at angles of 7° and 15°. This feature can be explained by comparison of the actual mass consumption of the electrodes after 6 h of NP synthesis with the NP mass-size distribution (Figure 8). Additionally, with an increase in the angle between electrodes' end faces, an increase in the energy efficiency of the NP synthesis is evident.

3.2. The Influence of the Degree of Electrodes Asymmetry on the NP Synthesis with the UDP

The UDP, which occurs when the ballast resistor is inserted in the discharge circuit, was characterized by approximately twice less amplitude and duration comparable to one half-wave of the ODP. Thus, the total electric charge passing through the spark gap in a single discharge pulse was about 7 times less for the UDP than for the ODP. For this reason, the electrical erosion wear of the electrodes during the several hours of the experiment was not sufficient to form on the cathode a significant area parallel to the anode, even at the angle of 10°. Therefore, experiments on the electrical erosion of electrodes and NP synthesis by the UDP were carried out with unchanged electrode shape at the macroscale.

It was found that, in processes of NP synthesis with the UDP, in cases of parallel arrangement of electrodes' ends or with a small slope of the cathode surface (3°), an electroerosive instability of the spark process occurred. It was manifested in the localization of spark channels to some response areas of electrodes end faces and was random in time. On the anode end face, a protrusion from the electrodes material had been appearing at the location of the spark channel. The protrusion grew over time and reduced the gap between electrodes. Reaching the cathode, it formed a metal bridge between electrodes, short-circuiting the interelectrode gap and ceasing spark discharges and NP synthesis. The start of the protrusion's rapid growth from the anode to the cathode is clearly indicated by the sudden change in functions of the capacitor charging voltage and the frequency of the capacitor discharges. This is represented in Figure 9 by black and red lines. The time of the beginning of the metal protrusion formation on the anode is longer for the cathode end face orientation at the angle of 3° than for the case of parallel arrangement of electrodes' ends. In the case of the cathode end face oriented at an angle, NP deposition on the anode reduces as it becomes easier for NPs to leave the interelectrode gap. Obviously, the protrusion is formed by NPs synthesized from the cathode material and transferred to the anode.

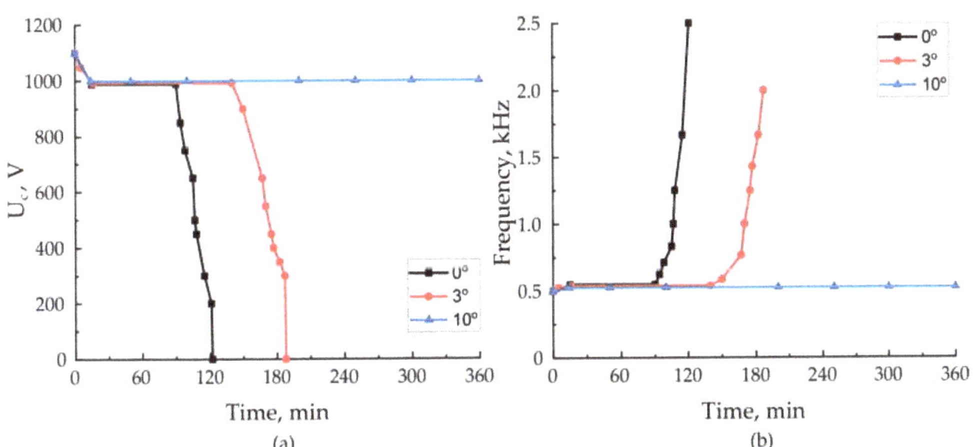

Figure 9. The dependences of (**a**) the capacitor charging voltage and (**b**) the frequency of capacitor discharges for the case of synthesis with the UDP for electrodes' ends arranged in parallel and at the angles of 3° and 10°.

The time dependences of modal sizes and concentrations of NP agglomerates synthesized in the SD with the UDP at different arrangements of the electrode ends are shown in Figure 10. From the comparison of these dependences, one can see an increase in the initial modal size of agglomerates with an increase in the angle between electrodes' end faces. This can be explained in the analogy with the case of the ODP by the localization of SDs in the area of the shortest distance between electrodes that leads to the increased current density in the SD and to the synthesis of large NPs.

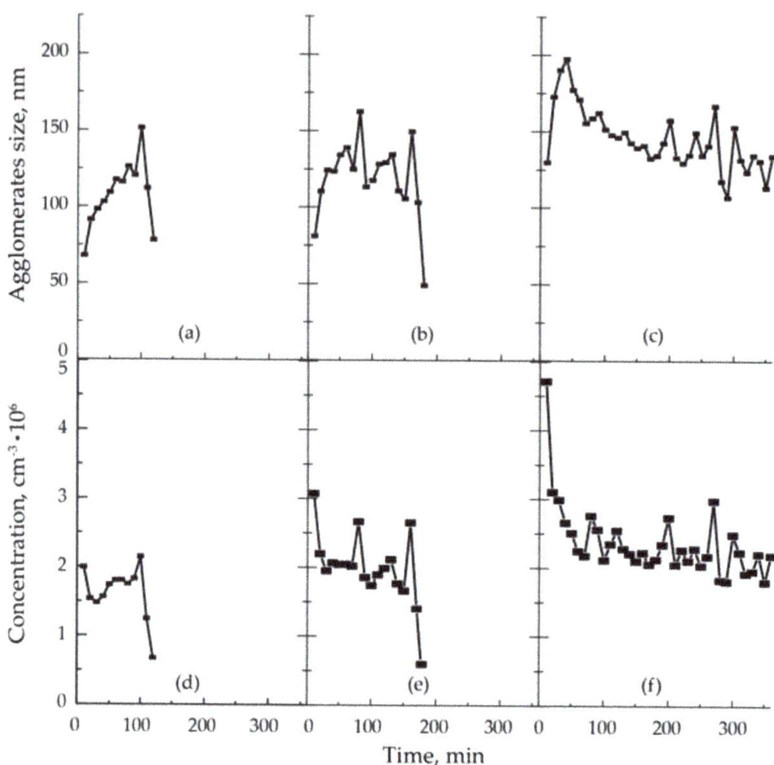

Figure 10. Time dependences of the modal size of NP agglomerates and their concentrations during the electrodes wear by the UDP for electrodes ends arranged (**a**,**d**) in parallel and at the angles of (**b**,**e**) 3° and (**c**,**f**) 10°.

Since, in the case of the UDP, a significant part of the capacitor's energy was dissipated on the ballast resistor, the energy spent on the electrical erosion of the electrode material was much lower than in the case of the ODP, and the synthesized NP agglomerates had smaller dimensions. Herewith, for the arrangement of electrodes ends at the maximum angle (10°), an asymptotic approximation of the dependence of modal sizes of NP agglomerates from time to the average size of about 130 nm is observed.

In the process of NP synthesis with the UDP, the anode end face is covering with an array of dendrite-like NPs. The cathode end face is modified by electrical erosion, followed by the formation of a set of overlapping microcraters. Such microcraters are formed as a result of evaporation and spilling of metal heated by the SD at places where it occurs on the cathode. According to the SEM images of electrode ends shown in Figure 11, one can observe the structure of their typical surface after the prolonged treatment by SDs. Since craters are found only on the cathode, we can logically assume that the anode is covered with NPs synthesized on the cathode. Thus, NPs synthesized from the cathode material are partially transferred to the anode and partially carried away in the form of aerosol by the transport gas. Herewith, the average size of craters turns out to be much smaller for the UDP than for the ODP (Figure 7c,d and Figure 11c,d). Moreover, the average crater's dimensions were found to be larger at the area of spark channel location on the cathode (Figure 11c) opposite the growing protrusion on the anode than at the cathode area aside the spark channel location (Figure 11d).

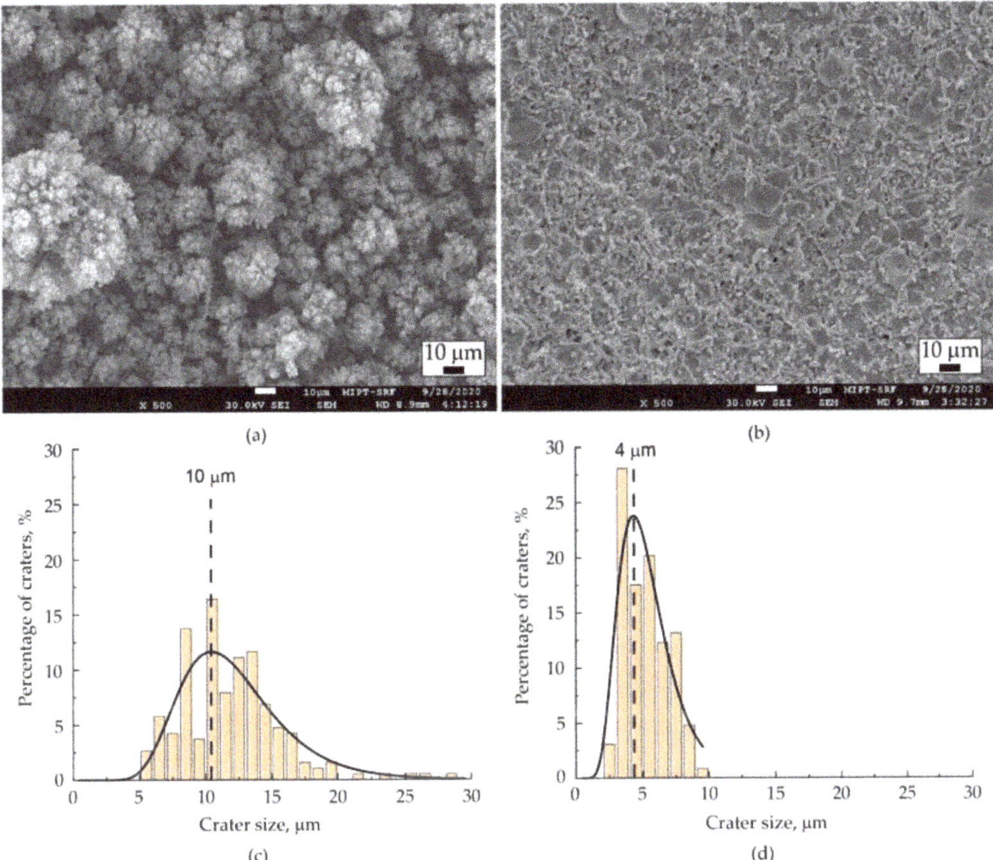

Figure 11. SEM images of typical surfaces of the (**a**) anode and (**b**) cathode after prolonged processing in the SD with the UDP and size distributions of craters on the cathode (**c**) at the area of spark channel location opposite the growing protrusion on the anode and (**d**) at the area aside the spark channel location for the case of the initial parallel arrangement of electrodes' ends.

A protrusion formed on the anode with the UDP was studied in optical and scanning electron microscopes (Figure 12). The photo in Figure 12a shows the localization of the spark channel in the discharge process at the stage of growth of the protrusion from the upper electrode (anode). Moreover, before the main protrusion growth begins, many clusters of NPs are gather on the anode in the form of micro-protrusions, which are located in a ring at a distance of about 2 mm from the center of the anode (in Figure 12b inside the dashed circle). When one of those protrusions begins to dominate, the growth of remaining protrusions stops, except for a single one quickly moving to the cathode, closing the interelectrode gap with a conducting bridge. Localization of the SD channel from a growing protrusion is similar to the previously observed SD processes with the pin-to-plane electrode configuration for pin-cathode [37] and pin-anode cases [38].

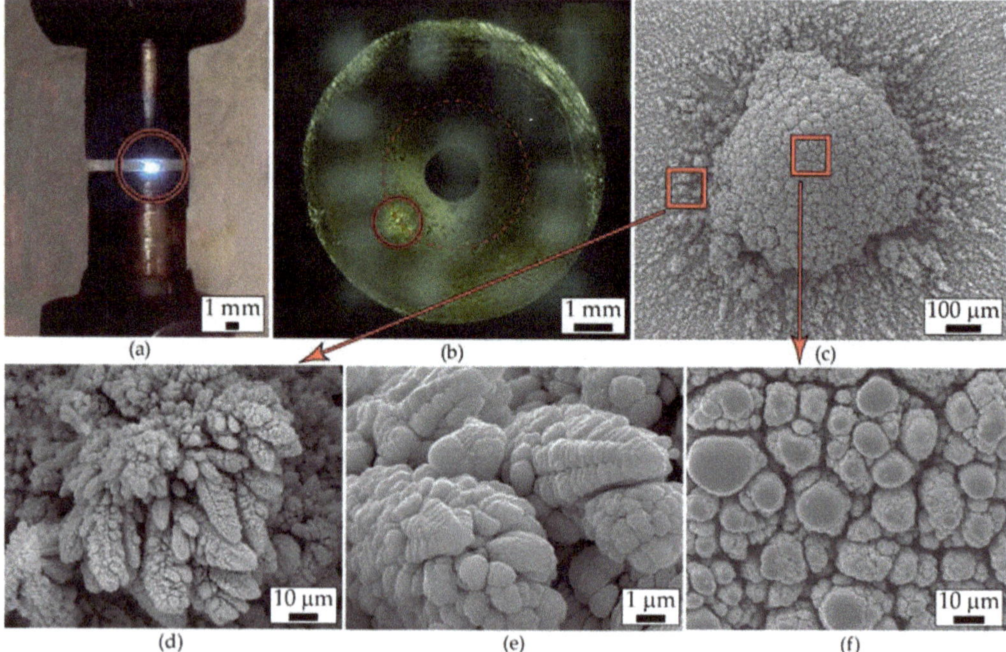

Figure 12. (**a**) Photo of the SD localization in the spark gap at the point of growth of the protrusion from the anode; (**b**) Optical microscope image of the anode end face with the designation of micro-protrusions formation area (dashed circle) and the dominant protrusion (solid circle); (**c**–**f**) SEM images of characteristic parts of the growing protrusion microstructure. Subfigure (**e**) is a magnified part of subfigure (**d**).

The microstructure study of the growing conductive protrusion (Figure 12c) in SEM allows us to conclude that in the central part of the protrusion the stacking of dendrite-like NP agglomerates is dense (Figure 12f) and on its periphery, the NP agglomerates assemble in the form of clusters with a loose stacking (Figure 12d,e). At the same time, many dendrite-like agglomerates in the central part of the protrusion are molten, probably due to their participation in the transfer of the SD current to the anode (Figure 12f).

In experiments with the UDP, we analyzed the dispersed composition of aerosol NPs by TEM and measured the mass loss of the cathode and anode. According to a series of TEM images of NPs (for example, Figure 13a–c) for each of the three variants of the electrodes' ends arrangement, the mass-size distributions of NPs (Figure 13d–f) were determined. It indicates that with an increase in the degree of electrodes asymmetry the NP agglomeration increases. In the case of electrodes parallel ends, a large number of isolated NPs were observed (Figure 13a). In the case of the electrodes' ends arrangement at the angle of 3°, the particles mainly formed small agglomerates. At the angle between ends of 10°, the energy release density in the discharge became sufficient for the fusion of NPs with the necks' formation specific for strong aggregation. At the same time, in all cases, a great number of NPs with sizes of 2–5 nm were observed, which did not make a significant contribution to the mass distribution. However, we have not detected NPs larger than 40 nm in these experiments, probably due to an order of magnitude lower energy release in the spark gap compared to similar experiments with the ODP.

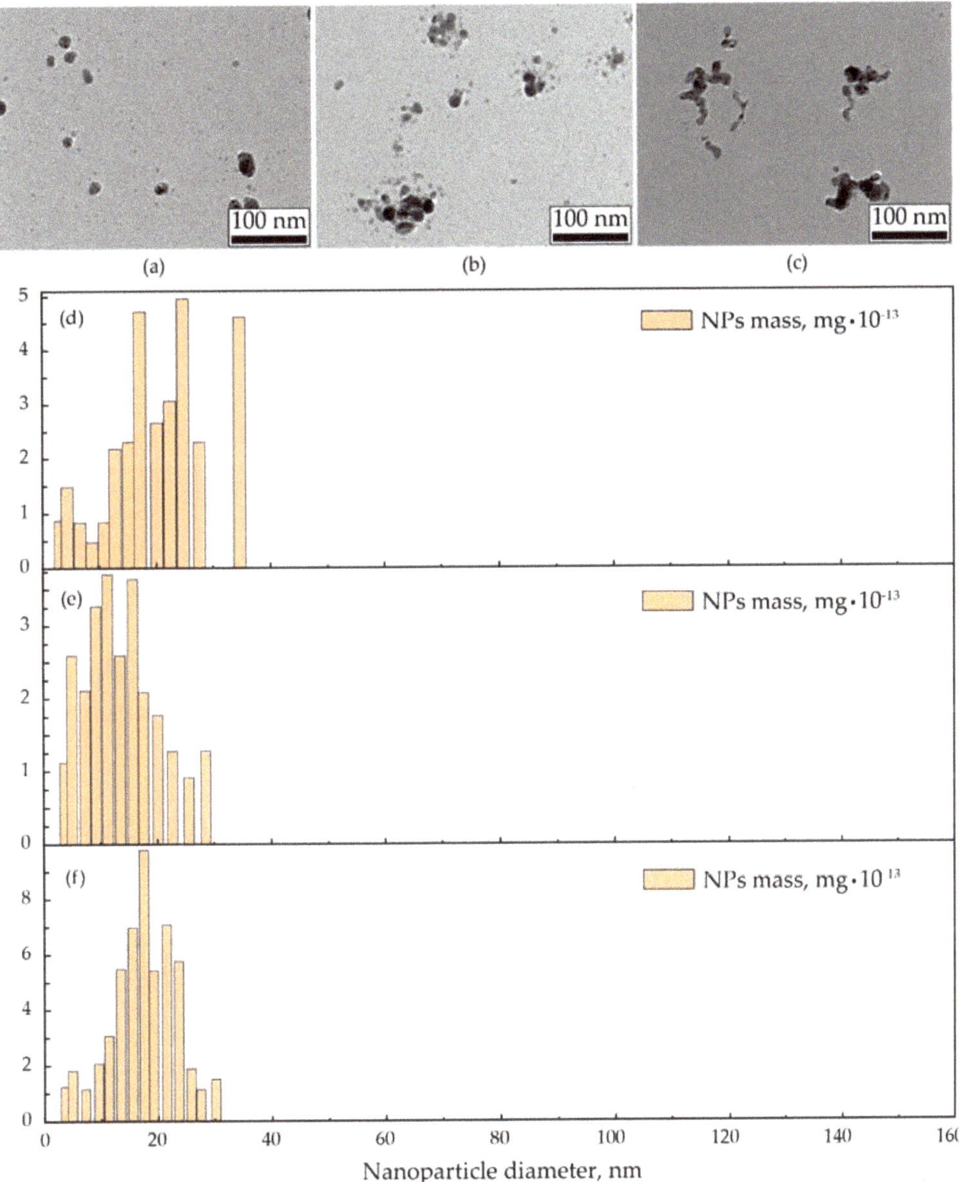

Figure 13. (**a**–**c**) TEM images and (**d**–**f**) mass-size NPs distributions for the cases of NP synthesis with the UDP for electrodes' end faces arranged (**a**,**d**) in parallel and at angles (**b**,**e**) 3° and (**c**,**f**) 10°.

Analysis of changes in the electrodes' mass after the completion of the SD process with the UDP showed that the main mass loss was at the cathode, and the anode mass might even slightly increase (Table 2). The increase in mass is logically associated with the transfer of synthesized NPs to the anode, including the growth of a metal protrusion leading to the formation of an interelectrode bridge. From Table 2, one can see that the electrodes' mass loss (mainly for the cathode) grows with an increase in the degree of electrode asymmetry. However, the energy efficiency in the case of the UDP varies slightly depending on the angle between electrodes' ends.

Table 2. The electrodes' mass loss, the energy release in the interelectrode gap, and the energy efficiency of the NP synthesis with the UDP.

Angle, °	ΔM_{anode}, mg	$\Delta M_{cathode}$, mg	ΔM, mg	E_t^d, kJ	m_e^d, $\frac{\mu g}{kJ}$
0	-0.68 ± 0.01	3.41 ± 0.01	2.73 ± 0.01	18.0 ± 0.2	152 ± 2
3	0.33 ± 0.01	5.01 ± 0.01	5.34 ± 0.01	28.7 ± 0.3	186 ± 2
10	-0.14 ± 0.01	7.72 ± 0.01	7.58 ± 0.01	60.9 ± 0.6	124 ± 1

4. Discussion

As established in the experiments presented above with the pulse-periodic SD, in the regimes with the ODP and the electrodes' asymmetric arrangement during the NP synthesis, their protruding parts are increasingly worn out by electrical erosion, forming parallel surfaces that increase in area and are covered with craters from spot discharges. At each moment of time, a certain area of the electrode is involved in electric discharge processes, usually in the form of a segment of the electrode cross-section (for example, in Figure 4b). Thus, we can say that from the start of the pulse-periodic SD process, equivalent working areas of electrodes, in the case of their ends arranged at an angle, increase from a minor value to the full cross-sectional area of the electrode perpendicular to its axis.

The kinetics of the increase in the working area of the electrodes arranged at an angle is correctly displayed by the observed patterns of decreasing time dependencies of the average NP agglomerate size and their concentration measured in the flow by NP electrical mobility (Figure 5). Additionally, the role of the small working area of electrodes initially arranged at an angle is in the greater average energy efficiency m_e^s over 6 h of the process for higher angles between electrodes end faces (Table 1). The average energy efficiency is the integral with respect to time from current energy efficiency function, monotonically decreasing from the start highest value. Experimental data obtained allow us to estimate the decrease in kinetics of the current energy efficiency and its maximum initial value. The current energy efficiency

$$m_e(t) = \mu(t)/P_d(t) \tag{6}$$

is followed from (5) and expressed as the mass loss rate by synthesized NPs $\mu(t)$ per current power released in the interelectrode gap determined from (2) and (4):

$$P_d(t) = \frac{R_s}{R_e}\frac{CU_c^2(t)}{2}\nu(t). \tag{7}$$

The mass loss rate by synthesized NPs $\mu(t)$ can be estimated from experimental dependencies of the concentration $N(t)$, the average size of NP agglomerates $D(t)$, measured by their electrical mobility (Figure 5), and from the value of the electrode mass loss ΔM (Table 1) [39]:

$$\mu(t) = \frac{dM}{dt} = A\frac{\pi}{6}D^3(t)\rho N(t)V = \Delta M\,\frac{N(t)D^3(t)}{\int_0^T N(t)D^3(t)dt}, \tag{8}$$

where ρ is the density of the NP material, V is the gas flow through the synthesis chamber and A is the constant coefficient that takes into account the analyzed fraction of the aerosol flow and the difference between the mass-weighted average NP size and the average size in terms of electrical mobility, which are reduced in the final expression. The calculations also use the expression for the total electrode mass loss during the time T as the integral from the mass loss rate by the synthesized NPs:

$$\Delta M = \int_0^T \mu(t)dt = A\rho V\frac{\pi}{6}\int_0^T N(t)D^3(t)dt. \quad (9)$$

Dependences $\mu(t)$, $m_e(t)$ and $P_d(t)$ calculated from experimental data for the three cases of the degree of electrode asymmetry are shown in Figure 14. The electrical power $P_d(t)$ entering the spark gap varies slightly with time except for the first 5 min, during which the working area is heated by SDs. However, the mass loss rate and the energy efficiency are described by steeply and monotonically decreasing functions.

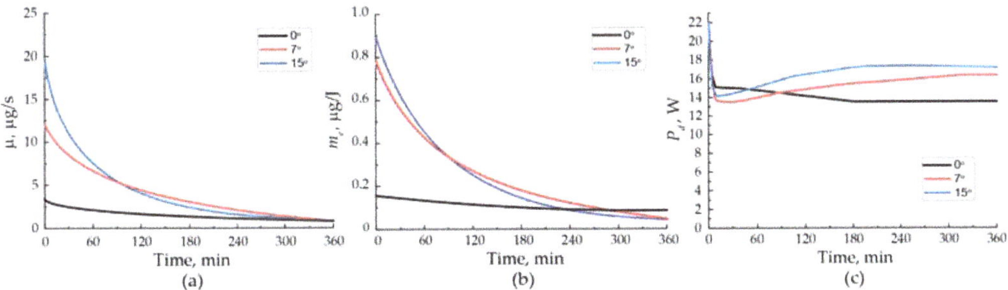

Figure 14. (**a**) The electrodes' mass loss rate, (**b**) energy efficiency, and (**c**) power released in the spark gap for the three cases of the degree of the electrode asymmetry with the ODP.

The observed effect of the energy efficiency decreasing or of the increasing in the equivalent working area of electrodes' ends initially arranged at an angle correlates qualitatively with the theoretically predicted effect of reducing the energy efficiency of NP synthesis with an increase in the diameter of the electrodes used [21]. Previously, it was noted that a higher mean temperature of electrodes increases mass production [23,36,40] and can lead to an increased production of large particles. This is realized in our experiments at the initial moments of the discharge process between electrodes ends arranged at an angle, when the equivalent working area of electrodes is small and the average density of the discharge current is high.

In experiments with the UDP, even when a cathode end had the angle of 10°, the synthesis of parasitic particles (more than 40 nm) was not observed, and the agglomeration of NPs was not significant. This occurred because, in the case of the UDP, the energy released in the spark gap per pulse was almost an order of magnitude less than in the case of the ODP. Therefore, there was no high energy release density even when the discharge was localized between beveled electrodes.

NP arrays on electrodes' end surfaces, as shown in Figure 4e, appeared in the case of an asymmetric arrangement of electrode end faces with the ODP. These surfaces, covered with NPs and not covered with discharge craters, indicate the deposition of a significant part of synthesized NPs back to the electrode and, probably, the transfer from the opposite electrode. Earlier in [23,36], the authors also noted the transfer of large particles to the opposite electrode and the transfer of NPs synthesized from the cathode material on the cathode [41,42].

Our experiments on the NP synthesis by the SD with the UDP, in which the mass of the anode increased (Table 2), clearly indicate the transfer of a significant proportion of synthesized NPs from the cathode to the anode. It is known that in the experiments with the UDP, each electrode works mainly as anode or as cathode and the electrical erosion of the cathode material turns out much higher than that of the anode, which was more accurately measured for larger interelectrode gaps in [43,44]. The width of the interelectrode gap of about 1 mm in our work contributed to an increased transfer of synthesized NPs to the opposite electrode. Since the erosion wear of the anode is much less than that of the cathode under these conditions, the material transferred from the cathode compensates the mass loss of the anode and can even increase it.

The origin of protrusions of NP clusters at the anode occurs due to the increased transfer of synthesized NPs from the cathode. These protrusions contribute to the electric field localization and, consequently, increase the probability of an electric breakdown of the gas in front of such protrusions. The growth of such protrusions is unstable, since the highest of them will more likely increase the local electric field and cause subsequent SD from the cathode in its direction. It is this single protrusion that grows and localizes the entire discharge process from the opposite region of the cathode. The growth of such a protrusion can be considered as the unstable process of the SD developing with a positive feedback. Moreover, the discharge process is stopped when the growing protrusion reaches the cathode and forms a conductive metal bridge between electrodes. To avoid instability due to the protrusion growth, it is necessary to create conditions for the removal of synthesized NPs from the interelectrode gap with higher efficiency. To do this, the rate of removal of NPs from the interelectrode gap must be greater than the rate of their synthesis.

Considering all the above, the reason for the absence of effects of the protrusions growth and the closure of the spark gap by a conducting bridge with the ODP becomes clear. In this case, the anode and cathode change their roles at each subsequent half-wave of the discharge current and are almost equal in the intensity of their electrical erosion.

5. Conclusions

The main result of this work is the revelation of another important parameter significantly affecting the processes of electrodes electrical erosion and the synthesis of NPs in a pulse-periodic SD—the degree of the electrodes' asymmetry in the rod-to-rod (rod-to-thick tube) configuration. It was demonstrated that it is possible to control the size and concentration of synthesized NPs by changing the degree of the electrodes' asymmetry by setting their end faces at a certain angle. With an increase in the degree of the electrodes' asymmetry, larger NPs (with sizes greater than 40 nm) appeared in the aerosol composition and their agglomeration increased. This behavior, observed with an increase in the angle between electrodes end faces, is associated with the localization of SD channels in the region of the shortest distance between electrodes and, consequently, with a local increase in the density of the discharge current and in the density of the power released in the discharge.

During the electroerosive wear of asymmetrically arranged electrodes by the ODP, a great part of the end faces of both electrodes became parallel. At the same time, parameters of the synthesized NPs approached those of the NPs synthesized on initially symmetric electrodes. This led to an asymptotic decrease in the average size of NP agglomerates to 150 nm for the case of the ODP and to 130 nm for the case of the UDP at the maximum degree of the electrodes' arrangement asymmetry.

In conclusion, it is also important to note that, with the ODP, an increase in the degree of electrodes' asymmetry led to an increased mass production rate and energy efficiency of NP synthesis, and a significant part of this increase was due to the contribution of large NPs with a size of more than 40 nm. Thus, in these regimes a large mass production rate was realized with a sharp deterioration in the qualitative composition of NPs. The parallel arrangement of electrodes' end faces provided the most stable mode of continuous synthesis of NPs of small sizes (less than 40 nm), both for the ODP and UDP. It is also important that, in all the cases, a significant proportion of synthesized NPs remained inside the interelectrode gap, settling on electrodes' end faces and not contributing to the energy efficiency of the synthesis. This effect can be minimized by controlling the SDG configuration and the speed parameters of blowing the electrode ends with gas, and by increasing the width of the gap between electrodes.

The effect of synthesized NP transfer to the opposite electrode with the UDP led to the occurrence of electroerosive instability. It appeared in the form of a protrusion on the anode surface, around which the SD process is localized, causing its further growth. This process led to an unstable regime of NP synthesis. As instability developed, the height of the protrusion on the surface of the anode increased, reducing the gap between electrodes. The protrusion grew due to the transfer of the cathode material by synthesized NPs to the area of its formation at the anode. Nevertheless, in the context of individual NP synthesis and the reduction of their agglomeration in the aerosol, the use of UDPs may be promising for practical applications, provided that the electroerosive instability development is prevented. Our further research will be devoted to finding conditions for preventing such instabilities.

Author Contributions: Conceptualization, K.K. and V.I.; Methodology, K.K.; Validation, K.K. and M.U.; Formal Analysis, K.K.; Investigation, K.K. and M.U.; Resources, V.I.; Data Curation, M.U., E.K. and A.L.; Writing—Original Draft Preparation, K.K.; Writing—Review & Editing, V.I.; Visualization, K.K. and M.U.; Supervision, A.E. and V.I.; Project Administration, V.I.; Funding Acquisition, V.I. and K.K. All authors have read and agreed to the published version of the manuscript.

Funding: This research was funded by RFBR and BRFFR, project number 20-53-00042 in part of the investigation of nanoparticle synthesis with the oscillation-damped discharge pulse and by RFBR, project number 19-33-90202 in part of the investigation of nanoparticle synthesis with the unipolar discharge pulse.

Conflicts of Interest: The authors declare no conflict of interest.

References

1. Schmidt-Ott, A. *Spark Ablation: Building Blocks for Nanotechnology*; CRC Press: Boca Raton, FL, USA, 2019; ISBN 978-1-00-073020-3.
2. Kruis, F.E.; Fissan, H.; Peled, A. Synthesis of Nanoparticles in the Gas Phase for Electronic, Optical and Magnetic Applications—A Review. *J. Aerosol Sci.* **1998**, *29*, 511–535. [CrossRef]
3. Scheeline, A.; Coleman, D.M. Direct Solids Elemental Analysis: Pulsed Plasma Sources. Available online: https://pubs.acs.org/doi/pdf/10.1021/ac00147a001 (accessed on 9 March 2021).
4. Mylnikov, D.; Efimov, A.; Ivanov, V. Measuring and Optimization of Energy Transfer to the Interelectrode Gaps during the Synthesis of Nanoparticles in a Spark Discharge. *Aerosol Sci. Technol.* **2019**, *53*, 1393–1403. [CrossRef]
5. Kohut, A.; Villy, L.P.; Ajtai, T.; Geretovszky, Z.; Galbács, G. The Effect of Circuit Resistance on the Particle Output of a Spark Discharge Nanoparticle Generator. *J. Aerosol Sci.* **2018**, *118*, 59–63. [CrossRef]
6. Evans, D.E.; Harrison, R.M.; Ayres, J.G. The Generation and Characterisation of Elemental Carbon Aerosols for Human Challenge Studies. *J. Aerosol Sci.* **2003**, *34*, 1023–1041. [CrossRef]
7. Byeon, J.H.; Park, J.H.; Hwang, J. Spark Generation of Monometallic and Bimetallic Aerosol Nanoparticles. *J. Aerosol Sci.* **2008**, *39*, 888–896. [CrossRef]
8. Vons, V.A.; de Smet, L.C.P.M.; Munao, D.; Evirgen, A.; Kelder, E.M.; Schmidt-Ott, A. Silicon Nanoparticles Produced by Spark Discharge. *J. Nanopart. Res.* **2011**, *13*, 4867. [CrossRef]
9. Lizunova, A.A.; Mylnikov, D.A.; Efimov, A.A.; Ivanov, V.V. Synthesis of Ge and Si Nanoparticles by Spark Discharge. *J. Phys. Conf. Ser.* **2017**, *917*, 032031. [CrossRef]
10. Mardanian, M.; Nevar, A.A.; Nedel'ko, M.; Tarasenko, N.V. Synthesis of Colloidal CuInSe2 Nanoparticles by Electrical Spark Discharge in Liquid. *Eur. Phys. J. D* **2013**, *67*, 208. [CrossRef]
11. Lee, D.-J. Highly Efficient Synthesis of Semiconductor Nanoparticles Using Spark Discharge. Ph.D. Thesis, Graduate School, Seoul National University, Seoul, Korea, 2018.
12. Tseng, K.-H.; Chang, C.-Y.; Chung, M.-Y.; Cheng, T.-S. Fabricating TiO2nanocolloids by Electric Spark Discharge Method at Normal Temperature and Pressure. *Nanotechnology* **2017**, *28*, 465701. [CrossRef]
13. Takao, Y.; Awano, M.; Kuwahara, Y.; Murase, Y. Preparation of Oxide Superconductive Composite by an Electrostatic Mixing Process. *Sens. Actuators B Chem.* **1996**, *31*, 131–133. [CrossRef]
14. Chadda, S.; Ward, T.L.; Carim, A.; Kodas, T.T.; Ott, K.; Kroeger, D. Synthesis of $YBa_2Cu_3O_{7-\gamma}$ and $YBa_2Cu_4O_8$ by Aerosol Decomposition. *J. Aerosol Sci.* **1991**, *22*, 601–616. [CrossRef]
15. Efimov, A.A.; Potapov, G.N.; Nisan, A.V.; Ivanov, V.V. Controlled Focusing of Silver Nanoparticles Beam to Form the Microstructures on Substrates. *Results Phys.* **2017**, *7*, 440–443. [CrossRef]
16. Volkening, F.A.; Naidoo, M.N.; Candela, G.A.; Holtz, R.L.; Provenzano, V. Characterization of Nanocrystalline Palladium for Solid State Gas Sensor Applications. *Nanostructured Mater.* **1995**, *5*, 373–382. [CrossRef]
17. Ivanov, V.V.; Efimov, A.A.; Myl'nikov, D.A.; Lizunova, A.A. Synthesis of Nanoparticles in a Pulsed-Periodic Gas Discharge and Their Potential Applications. *Russ. J. Phys. Chem.* **2018**, *92*, 607–612. [CrossRef]

18. Mohammed, A.M. Fabrication and Characterization of Gold Nano Particles for DNA Biosensor Applications. *Chin. Chem. Lett.* **2016**, *27*, 801–806. [CrossRef]
19. Khabarov, K.M.; Kornyushin, D.V.; Masnaviev, B.I.; Tuzhilin, D.N.; Efimov, A.A.; Saprykin, D.L.; Ivanov, V.V. Laser Sintering of Silver Nanoparticles Deposited by Dry Aerosol Printing. *J. Phys. Conf. Ser.* **2019**, *1410*, 012217. [CrossRef]
20. Khabarov, K.; Kornyushin, D.; Masnaviev, B.; Tuzhilin, D.; Saprykin, D.; Efimov, A.; Ivanov, V. The Influence of Laser Sintering Modes on the Conductivity and Microstructure of Silver Nanoparticle Arrays Formed by Dry Aerosol Printing. *Appl. Sci.* **2020**, *10*, 246. [CrossRef]
21. Domaschke, M.; Schmidt, M.; Peukert, W. A Model for the Particle Mass Yield in the Aerosol Synthesis of Ultrafine Monometallic Nanoparticles by Spark Ablation. *J. Aerosol Sci.* **2018**, *126*, 133–142. [CrossRef]
22. Feng, J.; Biskos, G.; Schmidt-Ott, A. Toward Industrial Scale Synthesis of Ultrapure Singlet Nanoparticles with Controllable Sizes in a Continuous Gas-Phase Process. *Sci. Rep.* **2015**, *5*, 15788. [CrossRef]
23. Tabrizi, N.S.; Ullmann, M.; Vons, V.A.; Lafont, U.; Schmidt-Ott, A. Generation of Nanoparticles by Spark Discharge. *J. Nanopart. Res.* **2008**, *11*, 315. [CrossRef]
24. Ivanov, V.V.; Efimov, A.A.; Mylnikov, D.A.; Lizunova, A.A.; Bagazeev, A.V.; Beketov, I.V.; Shcherbinin, S.V. High-Efficiency Synthesis of Nanoparticles in a Repetitive Multigap Spark Discharge Generator. *Tech. Phys. Lett.* **2016**, *42*, 876–878. [CrossRef]
25. Lizunova, A.; Mazharenko, A.; Masnaviev, B.; Khramov, E.; Efimov, A.; Ramanenka, A.; Shuklov, I.; Ivanov, V. Effects of Temperature on the Morphology and Optical Properties of Spark Discharge Germanium Nanoparticles. *Materials* **2020**, *13*, 4431. [CrossRef]
26. Yang, F.; Bellotti, M.; Hua, H.; Yang, J.; Qian, J.; Reynaerts, D. Experimental Analysis of Normal Spark Discharge Voltage and Current with a RC-Type Generator in Micro-EDM. *Int. J. Adv. Manuf. Technol.* **2018**, *96*, 2963–2972. [CrossRef]
27. D'Urso, G.; Maccarini, G.; Quarto, M.; Ravasio, C.; Caldara, M. Micro-Electro Discharge Machining Drilling of Stainless Steel with Copper Electrode: The Influence of Process Parameters and Electrode Size. *Adv. Mech. Eng.* **2016**, *8*. [CrossRef]
28. Ahmad, S.; Laiho, P.; Zhang, Q.; Jiang, H.; Hussain, A.; Liao, Y.; Ding, E.-X.; Wei, N.; Kauppinen, E.I. Gas Phase Synthesis of Metallic and Bimetallic Catalyst Nanoparticles by Rod-to-Tube Type Spark Discharge Generator. *J. Aerosol Sci.* **2018**, *123*, 208–218. [CrossRef]
29. Han, K.; Kim, W.; Yu, J.; Lee, J.; Lee, H.; Gyu Woo, C.; Choi, M. A Study of Pin-to-Plate Type Spark Discharge Generator for Producing Unagglomerated Nanoaerosols. *J. Aerosol Sci.* **2012**, *52*, 80–88. [CrossRef]
30. Trad, M.; Nominé, A.; Tarasenka, N.; Ghanbaja, J.; Noël, C.; Tabbal, M.; Belmonte, T. Synthesis of Ag and Cd Nanoparticles by Nanosecond-Pulsed Discharge in Liquid Nitrogen. *Front. Chem. Sci. Eng.* **2019**, *13*, 360–368. [CrossRef]
31. Schwyn, S.; Garwin, E.; Schmidt-Ott, A. Aerosol Generation by Spark Discharge. *J. Aerosol Sci.* **1988**, *19*, 639–642. [CrossRef]
32. Megyeri, D.; Kohut, A.; Geretovszky, Z. Effect of Flow Geometry on the Nanoparticle Output of a Spark Discharge Generator. *J. Aerosol Sci.* **2021**, *154*, 105758. [CrossRef]
33. *Handbook of Physics*; Benenson, W.; Harris, J.W.; Stöcker, H.; Lutz, H. (Eds.) Springer: Berlin/Heidelberg, Germany, 2002; ISBN 978-0-387-95269-7.
34. Jüttner, B. Cathode Spots of Electric Arcs. *J. Phys. D Appl. Phys.* **2001**, *34*, R103–R123. [CrossRef]
35. Mesyats, G.A.; Bochkarev, M.B.; Petrov, A.A.; Barengolts, S.A. On the Mechanism of Operation of a Cathode Spot Cell in a Vacuum Arc. *Appl. Phys. Lett.* **2014**, *104*, 184101. [CrossRef]
36. Kohut, A.; Wagner, M.; Seipenbusch, M.; Geretovszky, Z.; Galbács, G. Surface Features and Energy Considerations Related to the Erosion Processes of Cu and Ni Electrodes in a Spark Discharge Nanoparticle Generator. *J. Aerosol Sci.* **2018**, *119*, 51–61. [CrossRef]
37. Parkevich, E.V.; Ivanenkov, G.V.; Medvedev, M.A.; Khirianova, A.I.; Selyukov, A.S.; Agafonov, A.V.; Mingaleev, A.R.; Shelkovenko, T.A.; Pikuz, S.A. Mechanisms Responsible for the Initiation of a Fast Breakdown in an Atmospheric Discharge. *Plasma Sources Sci. Technol.* **2018**, *27*, 11LT01. [CrossRef]
38. Parkevich, E.V.; Medvedev, M.A.; Khirianova, A.I.; Ivanenkov, G.V.; Selyukov, A.S.; Agafonov, A.V.; Shpakov, K.V.; Oginov, A.V. Extremely Fast Formation of Anode Spots in an Atmospheric Discharge Points to a Fundamental Ultrafast Breakdown Mechanism. *Plasma Sources Sci. Technol.* **2019**, *28*, 125007. [CrossRef]
39. Hinds, W.C.; Hinds, W.C. *Aerosol Technology: Properties, Behavior, and Measurement of Airborne Particles*; Wiley: Hoboken, NJ, USA, 1999; ISBN 978-0-471-19410-1.
40. Feng, J. Scalable Spark Ablation Synthesis of Nanoparticles: Fundamental Considerations and Application in Textile Nanofinishing. Ph.D. Thesis, Delft University of Technology, Delft, The Netherlands, 2016. [CrossRef]
41. Petrov, A.A.; Amirov, R.H.; Samoylov, I.S. On the Nature of Copper Cathode Erosion in Negative Corona Discharge. *IEEE Trans. Plasma Sci.* **2009**, *37*, 1146–1149. [CrossRef]
42. Amirov, R.H.; Petrov, A.A.; Samoylov, I.S. Nanoparticles Formation and Deposition in the Trichel Pulse Corona. *J. Phys. Conf. Ser.* **2013**, *418*, 012064. [CrossRef]
43. Cundall, C.M.; Craggs, J.D. Electrode Vapour Jets in Spark Discharges. *Spectrochim. Acta* **1955**, *7*, 149–164. [CrossRef]
44. Hemmi, R.; Yokomizu, Y.; Matsumura, T. Anode-Fall and Cathode-Fall Voltages of Air Arc in Atmosphere between Silver Electrodes. *J. Phys. D Appl. Phys.* **2003**, *36*, 1097–1106. [CrossRef]

Article

Impact of the Samples' Surface State on the Glow Discharge Stability in the Metals' Treatment and Welding Processes

Maksym Bolotov [1,*], Gennady Bolotov [1], Serhii Stepenko [2] and Pavlo Ihnatenko [3]

1 Department of Welding Technologies and Construction, Chernihiv Polytechnic National University, 95 Shevchenko Str., 14035 Chernihiv, Ukraine; bolotovgp@gmail.com
2 Department of Electrical Engineering and Information Measuring Technologies, Chernihiv Polytechnic National University, 95 Shevchenko Str., 14035 Chernihiv, Ukraine; serhii.stepenko@stu.cn.ua
3 Department of Machine Building and Wood Processing Technologies, Chernihiv Polytechnic National University, 95 Shevchenko Str., 14035 Chernihiv, Ukraine; piligrim.8394@gmail.com
* Correspondence: bolotovmg@gmail.com; Tel.: +380-633353906

Abstract: The low temperature plasma of glow discharge has found a widespread use as a heating source in welding and surface treatment of metals. The meticulous analysis of glow discharge's instabilities in these processes allowed us to highlight the physicochemical characteristics of the cathode surface (the welded or treated samples) as one of the main reasons of its transition into an electric arc—as a more stable form of gas discharges. The prolonged arc action on the samples surfaces inevitably leads to the disruption of the technological process and, consequently, to undesirable overheating of samples. In this regard, the main aim of this work is to study the influence of the macro- and micro relief of the cathode on the stable glow discharge existence in the processes of metals treatment and diffusion welding. It has been analytically established and experimentally supported that the glow discharge's stability is mainly affected by the sharp protrusions generated on the cathode surface because of samples pre-treatment by machining before welding. It has been established that the rough surface pre-treatment with the Rz about 60–80 μm decreases the pressure range of glow discharge sustainable existence from 1.33–13.3 kPa to 1.33–5.3 kPa compared with the surface machining with the Rz about 10 μm.

Keywords: diffusion welding; plasma; glow discharge; surface treatment; plasma techniques

Citation: Bolotov, M.; Bolotov, G.; Stepenko, S.; Ihnatenko, P. Impact of the Samples' Surface State on the Glow Discharge Stability in the Metals' Treatment and Welding Processes. *Appl. Sci.* **2021**, *11*, 1765. https://doi.org/10.3390/app11041765

Academic Editors: Alessandro Belardini and Mariusz Jasiński

Received: 10 December 2020
Accepted: 10 February 2021
Published: 17 February 2021

1. Introduction

Nowadays, to obtain qualitative permanent joints of heterogeneous materials the methods of welding in a solid state are widely used. The most prevalent of these is a diffusion bonding. The wide nomenclature of compounds creates a complex of specific requirements for diffusion bonding's energy sources. These requirements are mainly related to the acceptability of a wide range of materials and shapes of products, the accuracy of the specific heat capacity control and the ability of the wide regulation of the sample's temperature [1].

The distributed plasma of a glow discharge burning in a rarefied gas atmosphere at a pressure of 0.1–10 kPa is widely used in processes accompanied by the direct action of charged electric particles on the treated or welded materials. Known works consider the possibility of application of gas discharges technique in the field of thin films deposition, metals surfacing, treatment and modification of metals before welding [2–5]. Still, as practice has shown, diffusion welding [6] and thermal and chemical-thermal surface treatment [7] are the most appropriate. This is due to the high technological capabilities of the glow discharge, which in these processes can serve both a processing tool and a source of thermal energy for their implementation simultaneously. Additionally, a glow discharge has high technical, economic, and environmental indicators, for instance, high heating productivity, energy savings, and last but not least the absence of environmental pollution [8].

However, during surface treatment or welding of metals, the instabilities of samples temperature (discharge cathode in these processes), the pressure of the working gas, the voltage of the power supply, and a number of other factors, can lead to the transition of a glow discharge into another form of gas discharge—an electric arc [9] (Figure 1). In this case, the distributed glow discharge transforms into a contracted form. Its cathode spot narrows to a very small size, increasing the energy concentration dramatically. A prolonged action of a concentrated arc discharge inevitably leads to the local melting and destruction of samples.

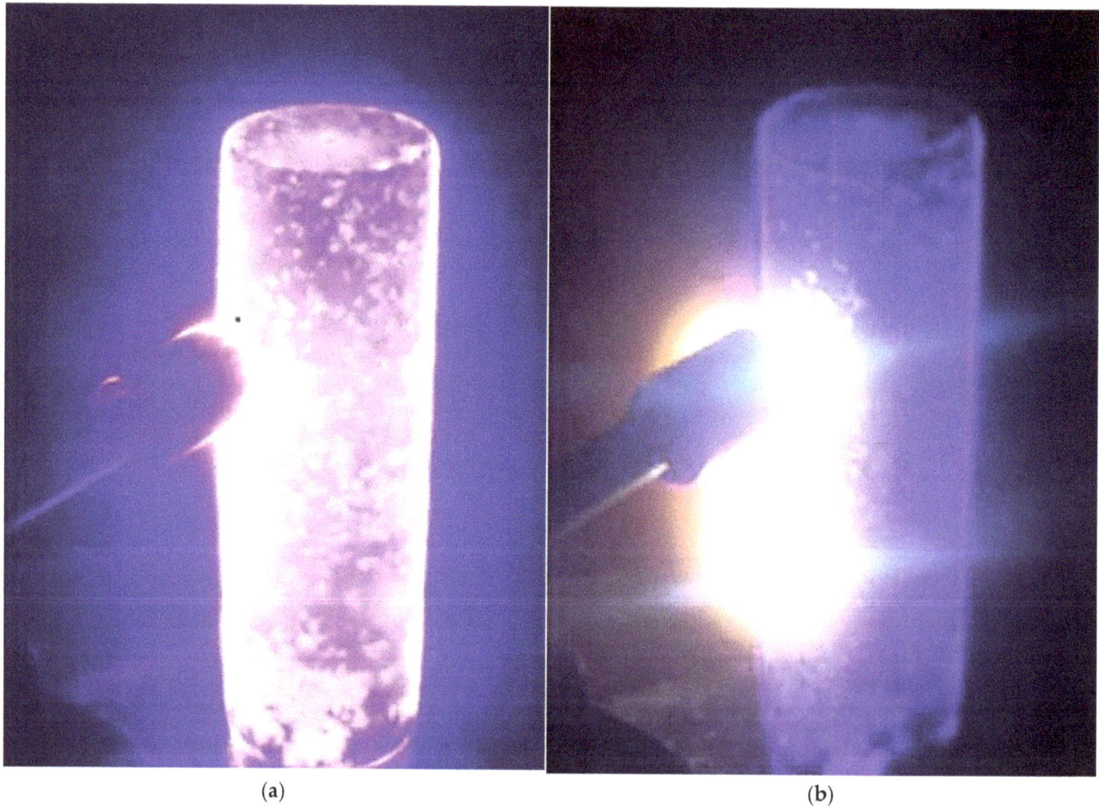

(a) (b)

Figure 1. The view of the glow discharge on the cathode surface in stable (**a**) and unstable (**b**) burning modes during the ion treatment (the bright spots to the right—arc breakdowns).

The probability of arcing increases with the rising of the total and specific power of the discharge. The energy characteristics of the ionic treatment processes are mainly determined by the magnitude of the discharge current and the gas pressure in the working chamber. The development of processes' productivity leads to the necessity of increasing their values. That, in turn, entails rising of average current density (j) in the cathode spot of the discharge and the average specific volumetric power (jE) in the discharge plasma [10]. However, with increasing of discharge energy characteristics in the interelectrode gap, the short-term local arc breakdowns with a duration of one or two half-periods of the rectified current can form (Figure 2).

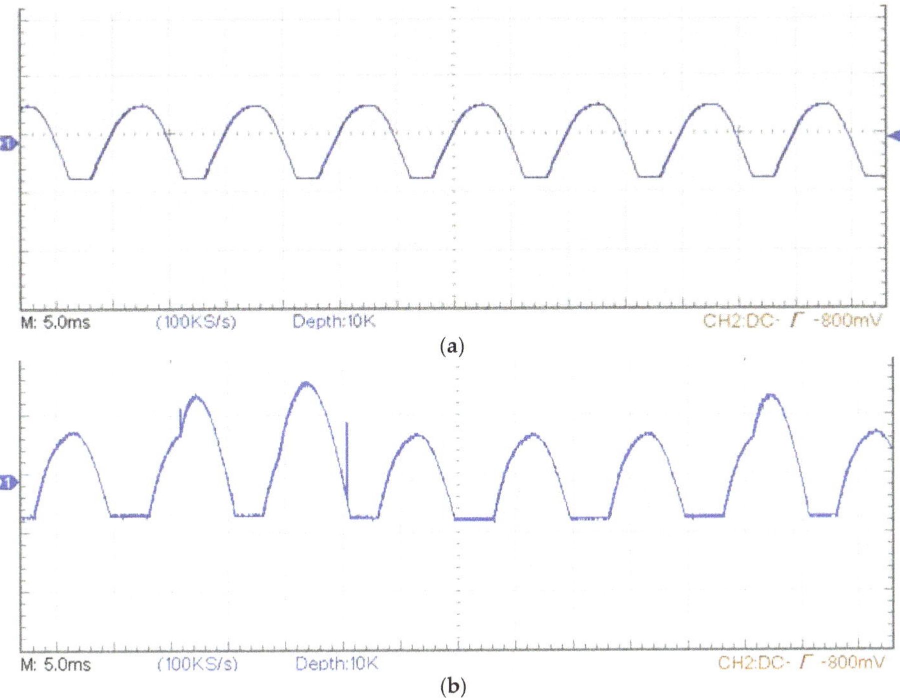

Figure 2. Oscillogram of the discharge current in stable (**a**) and unstable (**b**) mode. (Glow discharge current 4 A. Oscillograms were recorded from a shunt of 3 Ohm connected to the discharge circuit).

In [11,12], the glow discharge's instabilities are closely associated with its contraction (compression) and transition to a cord form of discharge with the increasing of volumetric power (jE) accordingly. The gas in the cord discharge heats up extremely while the burning voltage decreases and eventually the cord switches into an arc electric. Such a mechanism of glow discharge's instability is associated with volumetric processes in positive column of the discharge plasma, and therefore is more characteristic of extended discharges of the laser type [13], where the length of the interelectrode gap is 0.5–1.0 m. In the processes of ion treatment and welding of metals, a stationary DC glow discharge burns between the electrodes with a limited distance of 0.005–0.05 m. In these conditions, processes in the near-electrode regions of the discharge can affect negatively its stability. The largest voltage drop (100–300 V) and the greatest electric field strength accordingly are observed in the cathode region of the discharge where there are the processes of electron ionization and multiplication which determine the glow discharge existence. These processes are affected by conditions both in the volume of the cathode layer and on the surface of the cathode itself. They are quite fully investigated and described in [14–16]. However, they do not consider the impact of the cathode surface characteristics on the stability of the glow discharge while surface treatment and welding of metals, which makes it impossible to determine and establish their optimal values from the point of view of process productivity and discharge stability. In this regard, the aim of this work is to study the effect of the physicochemical characteristics of the samples surface on the stability of a glow discharge and the development, on this basis, of technological recommendations for the selection of values of the mode parameters that ensure the stable discharge existence while diffusion welding and metals treatment.

2. Methods

The main physicochemical characteristics of the metals surface includes its macro- and micro relief, as well as the presence of chemical compounds on it. Even after machining, the surface of the vast majority of structural metals is covered with a thin layer of natural oxide. Oxide films are not ideal dielectrics, but they possess a certain conductivity. The resistivity (ρ) of most oxide films of the processed metals (Fe, Cu, Mo, W, Ti, etc.) is 10^3–10^5 Ohm m, and the characteristic value of the dielectric constant is $\varepsilon \approx 2$–10 [17]. Therefore, from an electro-technical point of view, an oxide film's capacitance and resistance are connected in parallel. The capacitive properties of the film appear at times $t_0 \leq \varepsilon\varepsilon_0\rho \approx 3\ (10^{-8}$–$10^{-6})$ sec [18]. The maximum electric field strength due to the accumulation of charge on the oxide film over this time is defined as

$$E \approx \frac{1}{\varepsilon\varepsilon_0}\int_0^{t_0} jdt \approx jp, \qquad (1)$$

where j—the current density at the cathode of the discharge.

In a normal glow discharge, the gas pressure in the working chamber determines the current density in the cathode spot and at the pressures of 1.33–13.3 kPa it can reach 10^2–10^3 A/m^2. Thus, in a normal DC glow discharges at pressures characteristic of technological processes of metal treatment and welding, the electric field may not reach the values of the electric field strength breakdown for thin oxide films. As it was mentioned in [16], the heating of the oxide films to a 1200 K leads to a rapid decrease of its resistivity (four to six orders of magnitude). Therefore, taking into account the simultaneous heating of the films together with the samples and the noticeable decline in the film resistance with increasing of the temperature, the emergence of arcing breakdown is unlikely. Consequently, the presence of an oxide film on the surface of the cathode (product) can make a massive impact on the discharge's stability but only on the cold cathodes. With the heating of cathode, this factor becomes unimportant.

The next parameters characterizing the state of the cathode surface are its macro- and microrelief. In this case, the glow discharge's stability can be affected with the pronounced roughness protrusions obtained after machining, but not with the smoothly changing surface waviness. Figure 3 shows profile curves of the sample's surfaces obtained after treatment with a roughness of 60–80 μm (Figure 3a) and about 10 μm (Figure 3b), respectively. For this purpose, a profilometer TR—200 was used.

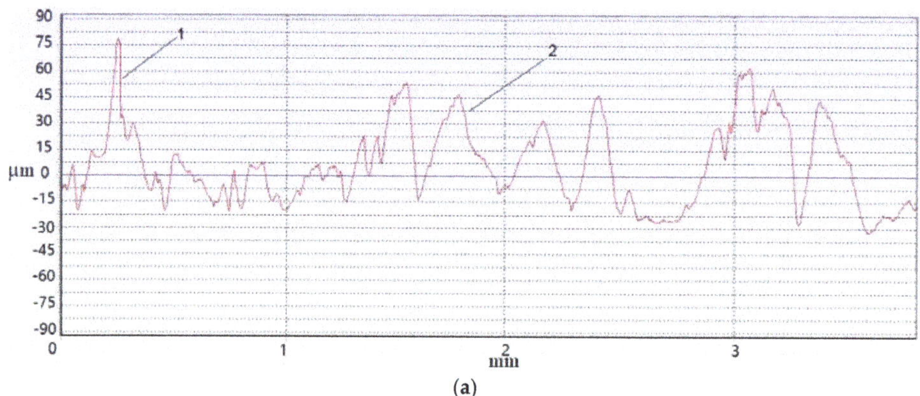

(a)

Figure 3. *Cont.*

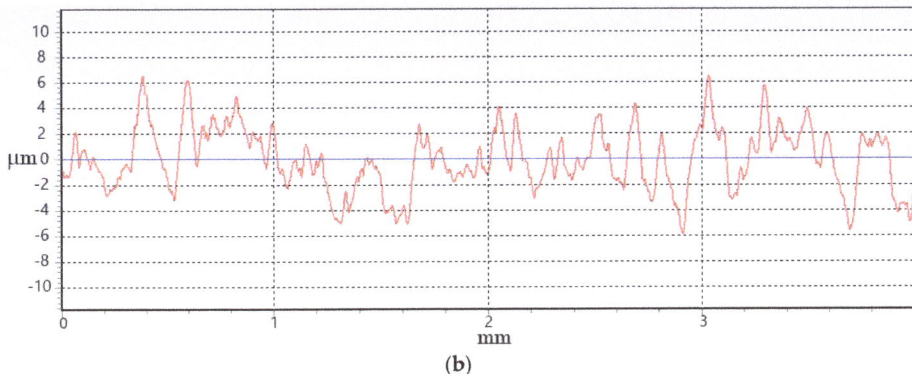

(b)

Figure 3. The profile curves of the samples surfaces obtained after machining by rotational turning with the roughness about 80 μm (**a**) and 10 μm (**b**): 1—single placed; 2—ridge-shaped protrusions.

The external view of the device and treated samples themselves are shown in Figure 4.

(a)

(b)

Figure 4. The view of a profilometer TR-200 (**a**) which is made up of: 1—display; 2—control panel; 3—sensor (pickup) and treated samples (**b**).

The undergoing intense ions bombardment, the protrusions heat up much more noticeably than the bulk of the electrode (cathode). This inevitably leads to the superheat point's emergence on the cathode surface to temperatures exceeding the melting or boiling point of the metal. In this case, a glow discharge can transit to the local vapor arcs [19,20]. This can be observed most likely in case when the surface of the protrusions is large enough to provide heating to high temperatures, and the heat transfer to the bulk of the samples is too small.

3. The Analytical Equation of a Glow Discharge Energy Stability while Treatment and Welding of Metals Depending on Cathode Surface State

Cylinders, cones or hemispheres can be taken as a model of individual micro protrusions [21]. The possibility of their heating to a boiling point is provided:

$$q_s = q_v - q_T, \tag{2}$$

where q_s—the energy perceived by the surface of the protrusion from ions bombarding the cathode: $q_s = SjU_ct$ (where S—the area of the lateral surface of the protrusion, U_c—the cathodic potential drop in the discharge, t—time); q_v—the heat content of the protrusion material at the boiling point: $q_v = Vc\gamma T_{boil}$ (where V—the protrusion volume, c—the heat capacity of the metal, and γ—the density); and q_T—the energy diverted from the protrusion into the sample: $q_T = 2\pi\lambda RT[1 - erf\left(\frac{R}{4at}\right)]^{-1}$, (where λ—the thermal conductivity of the cathode material, R—the radius of the protrusion base, and T—the cathode surface temperature).

In the processes of thermal ion treatment and diffusion welding, when the temperature of the samples is much lower than the boiling point of metal, condition (2) is feasible if $q_T \to 0$. In order to neglect the heat transfer to the samples, the value of t must be the same order as the glow discharge's transition time into the arc $t = 10^{-4}$–10^{-6} sec. So, then

$$q_s = q_v or SjU_ct = Vc\gamma T_{boil}, \tag{3}$$

In this case, for the micro relief of structural steels, it is necessary that the S/V ratio is 10^5–10^7 cm^{-1} (where S—the area of the lateral surface of the protrusion and V—the protrusion volume). Nevertheless, even at t = 1 sec, it is necessary that the value of $S/V \geq 10^2$ cm^{-1}. Such ratios of the surface and volume of the micro protrusion are possible only for thin protrusions and burrs. For real surfaces machined by turning or grinding, this ratio is much less than one, which indicates a low probability of fulfilling condition (2). Hence it follows that at the cathode's current densities corresponding to a glow discharge (up to 10^3 A/m^2), the probability of melting and evaporation of the roughness ridges is pretty small. This limits the possibility of thermionic emission from their vertices.

At the same time, sharp protrusions on the cathode surface creates the local distortions of the electric field nearby the cathode surface. Electric field distortions near the protrusions facilitates the attraction of positive ions toward them and as a result, the active points with the increased charges concentration are formed. This, in turn, leads to dramatic increase in the electric field strength in these regions. The length of this region (d_c) is determined by the cathode material as well as kind and pressure (p) of the working gas, and can be found from the ratio:

$$pd_c = c, \tag{4}$$

where c—a constant value for particular gas and its pressure (for a glow discharge burning in a nitrogen, c = 0.42 Pa m [22]). At a nitrogen pressure of 1.33–13.3 kPa, the average electric field strength near the discharge cathode is

$$E_{av} = \frac{U_c}{d_c}, \tag{5}$$

where U_c is the cathode potential drop, which is 215 V for a glow discharge in nitrogen.

In these conditions, the E_{av} reaches up to 10^6–10^7 V/m. A further increase in the electric field strength inevitably leads to the appearance of a current of field electron emission from the tip of the protrusion due to the ejection of electrons from the surface by a strong electric field. A noticeable field emission current, sufficient for the existence of an arc discharge, appears at the electric field strength of about 10^9 V/m. In this regard, it is advisable to assess the degree of influence of the protrusions surface roughness on a local increase in the electric field strength at the cathode of a glow discharge. Meanwhile, the magnitude of the field strength does not depend on the number of protrusions, but on their size and shape.

In mechanical engineering, the surface machining of samples of the third—sixth class of cleanliness is widely used. In this case, the maximum height of the surface protrusions varies, respectively, from Rz = 40–80 to 6–10 μm. Since the height of the micro protrusion is much less than the interelectrode gap (of 0.01 m or more), for calculation of electric field of such protrusion the electrostatics task has being solved about the conductive ellipsoid in an external field parallel to one of the main axes of the ellipsoid [23]. This is an equivalent to a semi-ellipsoidal protrusion on one of the flat electrodes, or to the gap between two electrodes parallel to each other (Figure 5) which exceeds protrusion height. If the x is the axis of the protrusion in the form of a semi-ellipsoid of rotation perpendicular to the electrode, where $x = 0$, then the field strength on the extension of this axis when x is greater than the height of the protrusion h is [22]:

$$E(x) = E_{av}\left[(1 - \frac{arth\frac{c}{x} - \frac{c}{x}}{arth\frac{c}{h} - \frac{c}{h}}) + \frac{1}{(arth\frac{c}{h} - \frac{c}{h})(\frac{x^2}{c^2} - 1)\frac{x}{c}}\right], \quad (6)$$

where E_{av} is the average electric field strength created in the cathode region of a glow discharge by a cathodic potential drop; c—half the distance between the ellipsoid focuses located on the x axis (Figure 5), determined, according to [24], as $c = \sqrt{a^2 – b^2}$.

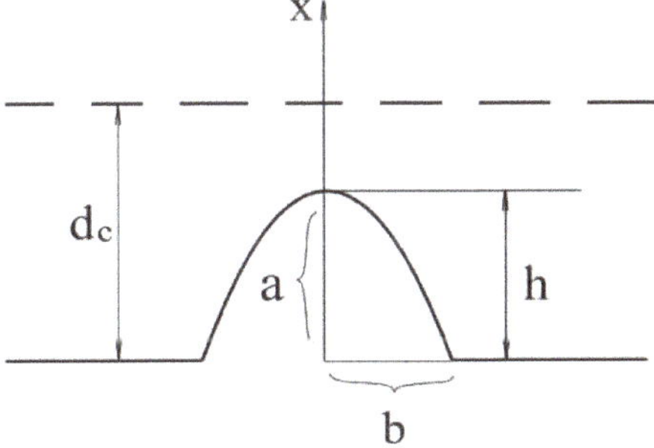

Figure 5. Scheme of a semi-elliptical protrusion on the surface of a flat cathode: a, b are the semi axes of the ellipsoid; h—the height of the protrusion; d_c—the length of the cathodic potential drop region.

The term in square brackets in this expression describes the field strength increasing due to the presence of a protrusion. Denote it by β_E, then the expression (6) can be written as $E(x) = E_{av}\ \beta_E$. The term in parentheses of expression (6) is always less than one and on the surface of the protrusion at $x = h$ vanishes. Therefore, the main is the second term, the value of which is maximum at minimum x, i.e., at $x = h$ and at the $x = d_c$ simultaneously. Figure 4 shows the dependence of the field enhancement β_E on the ratio h/d_c. Expression (6) describes an increase in the field strength on a flat smooth cathode from a protrusion having a smooth peak with a considerable radius. For a hemispherical protrusion having the same semi axes a = b, the field enhancement is $\beta_E \approx 3$ [25]. The results of calculating β_E from expression (6) for surface protrusions in the form of semi-ellipsoidal extended along the x-axis are shown in Figure 6. The graph also shows that, depending on the degree of protrusions elongation, the electric field enhancement at their apex can reach up to 10–30.

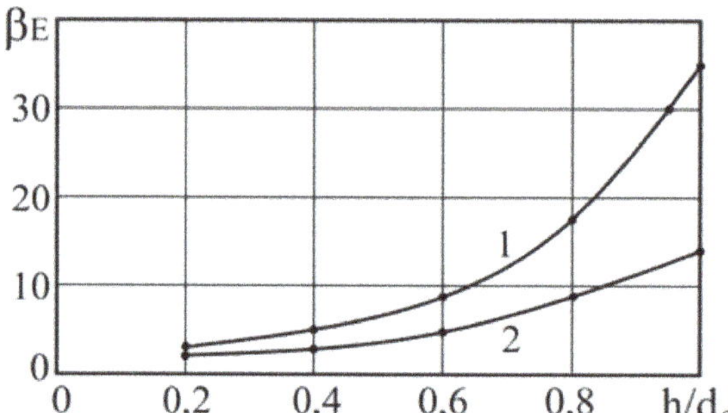

Figure 6. The dependence of the field enhancement on the cathode surface on the ratio of the protrusion's height and the length of the cathodic potential drop region: 1—ratio a/b = 4; 2—ratio a/b = 2.

In practice, after machining protrusions are formed on the surface, having an acute-angled shape, with a radius at the apex within the unit fractions of micrometers (Figure 3). The acute-angled protrusions contribute dramatically to a more intense electric field distortions and an additional increase in its intensity nearby the protrusion. Such an increase in electric field strength can be estimated by the coefficient [21]:

$$\mu = \frac{h}{r}, \tag{7}$$

where h—the height of the protrusion; r—the curvature radius on the peak of the protrusion.

For protrusions with a height of 20–40 μm or more with a radius of vertex curvature from fractions to units of micrometers (Figure 3), the value of μ can reach up to $\mu \geq 10$. Then the field strength near the vertices of such protrusions will be

$$E(x) = E_{av}\beta_E\mu, \tag{8}$$

Under conditions of increased gas pressure, the boundary of the region of cathodic potential drop declines sharply and approaches the vertices of micro protrusions roughness $x = d_c \approx h$. As a result, the local field strength near the vertices of such protrusions can reach $E(x) \geq 10^9$ V/m. Such values of the local electric field strength provide for a current density of field emission from the peak of the protrusions of $j_{fe} \geq 10^7$ A/m^2, sufficient for an arc breakdowns exciting in the interelectrode gap [26]. The latter is in many orders higher than the current density in the cathode spot of the glow discharge. It is suggested that in these conditions the Joule's heating and evaporation of the vertices of the protrusions entail the arc breakdowns [27]. In this case, condition (2) can already be fulfilled, i.e., the probability of melting and evaporation of protrusions increases dramatically. In turn, the latter contributes to the development of thermionic emission from these areas with the arc formation on the surface of the cathode spot of an arc discharge. As was mentioned above the long-term action of a stable arc discharge on a surface of the samples can lead to their melting and destruction.

Hence, for the prevention of a glow discharge transition into an electric arc, the length of the cathodic potential drop must exceed the maximum height of the roughness protrusions. The cleanliness class of the samples surface treatment determines this.

The adequacy of this assumption has been checked under conditions of ion treatment in a glow discharge in a nitrogen. The cylindrical c with dimensions of 20 × 60 mm made of steel A 659 CS Type 1020 (cathode of discharge simultaneously) were used. The glow discharge was powered from a controlled full-wave rectifier with an output voltage of 0–1000 V through a ballast resistor of 80 Ohm. The discharge current was of 4 A. A flat annular anode located at a distance of 0.008 m from the cathode surface was used as the second discharge electrode (Figure 7).

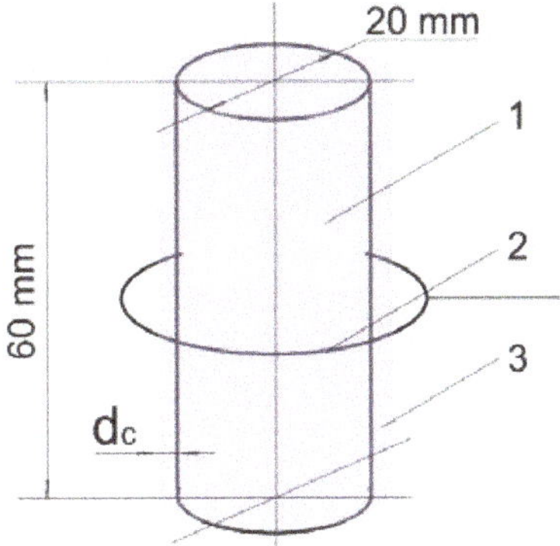

Figure 7. The schematic view of experiments: 1—treated samples (cathode); 2—anode ring; 3—plasma of glow discharge; d_c—the length of the cathode potential drop.

The steel samples obtained by rough, semi-finishing and finishing had the height of the surface roughness protrusions of 60–80, 30–40, and 10–15 μm accordingly. During treatment, a gradual increase in gas pressure in the working chamber was performed.

As the criteria of glow discharge stability, the limiting pressure at which the short-term arcs at the interelectrode gap emerge developing, apparently from the tops of the highest or most sharp protrusions was chosen. A further rising of a gas pressure is accompanied by an increase in the frequency of microarc discharges formation until a stable electric arc is established in the discharge gap. The cathode (products) temperature while treatment was 700–1000 K. The cathode's temperature was measured with the chromel-alumel thermocouple at three different points in the axial direction and then the results was averaged. The view of treatment by stable glow discharge and in the moments of arcing breakdowns (blurred by flare and multitude of micro-arcs) are shown in Figure 8.

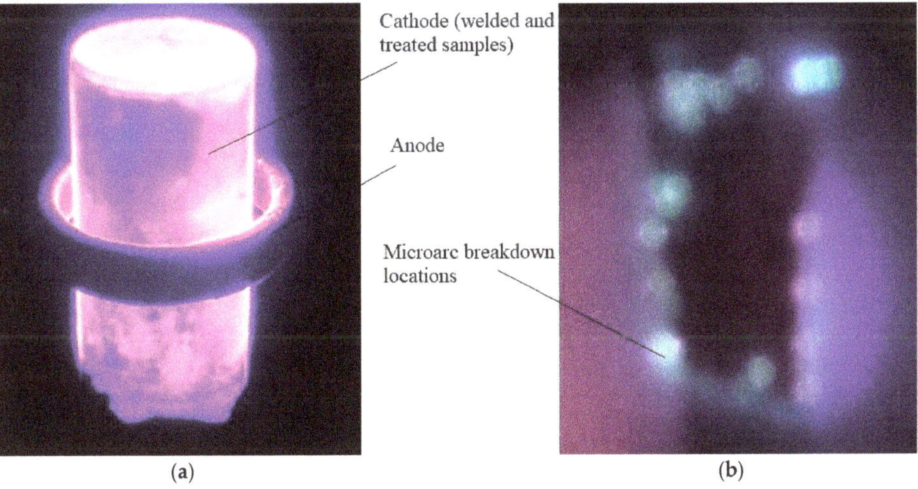

Figure 8. The visualized scheme of treatment in a stable (**a**) and disturbed (**b**) by a multiple micro-arcs breakdown glow discharge.

The experimental results are shown in Figure 9 in the form of a curve corresponding to the averaged over the results of a number of measurements, the limiting values of gas pressure depending on the parameters of the surface relief. Here along with the dimensions of the roughness protrusions, the values of the extent of the cathodic potential drop region corresponding to the gas pressures are given.

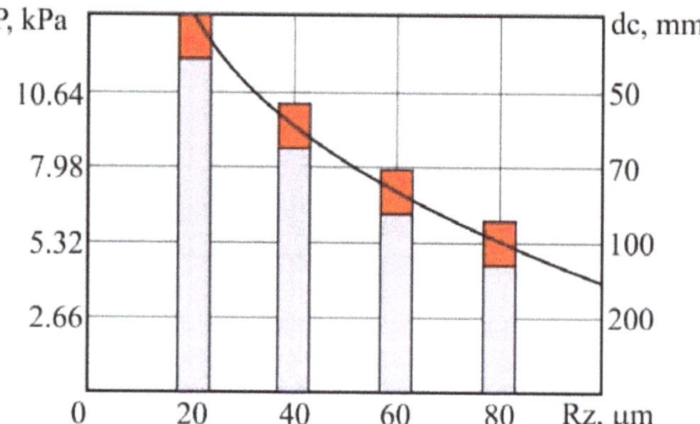

Figure 9. The experimental dependence of the working gas pressure (*P*) on the surface roughness parameters (Rz) according to the conditions of glow discharge stability: (d_c—cathodic potential drop region).

The obtained results indicate that the glow discharge's instabilities really begin to appear when the boundary of the cathodic potential drop region (d_c) approaches the vertices of the micro protrusions. A glow discharge remains stable at gas pressures below the obtained experimental curve. Knowing the samples surface characteristics in the form of the Rz = Rmax quantity, enables to determine the gas pressure boundary values for the various technological processes of ion treatment or diffusion bonding, as well, in advance.

4. Conclusions

It is shown that the low temperature plasma of DC glow discharge which burns in the active or inert gases at the medium pressures is the perspective source of surface heating in the processes of diffusion bonding and ion treatment of metals. At the same time, a number of factors on the cathode surface can emerge, which leads to the transition of a glow discharge into an electric arc as a more stable form of gas discharges. In this study, the impact of macro and micro relief of the samples on the stability of a glow discharge while diffusion bonding and treatment of metal has been analyzed and analytical equation of a glow discharge energy stability boundary has been obtained. The main conclusions of this study can be summarized, as follows:

1. The emergence of oxide films on the surfaces of specimens during welding or metals treatment does not lead to a significant disruption of a glow discharge stability as long as electrical field strength does not exceed the breakdown values.
2. On the other hand, the roughness of a cathode's surface affects the glow discharge stability at the working gas pressures when the height of the protrusions roughness becomes comparable to the length of the cathode potential drop region d_c.
3. The direct dependence of the d_c length on the gas pressure allows to determine the limit values of the latter based on the given characteristics of the samples surface microrelief which ensures the stable glow discharge existence during ion treatment and welding of metals. In our experiments the increase of the cathode surface roughness from 10–15 µm to 60–80 µm led to a rapid decrease of the region of the limiting

pressure of the stable glow discharge existence from 1.33–13.3 kPa to a 1.33–5.3 kPa, respectively.

Author Contributions: G.B.: conceptualization and methodology; M.B.: experiments, results validation, formal analysis, writing—original draft preparation; S.S.: literature review, main glow discharge's instabilities related with surface state of cathode, writing—review and editing; P.I.: investigation, data curation, getting profilograms. All authors have read and agreed to the published version of the manuscript.

Funding: This research is supported by the Ministry of Education and Science of Ukraine (Grant 0117U007259 "New high-tech energy-efficient heating source for precision welding, brazing and surface treatment of metals").

Conflicts of Interest: The authors declare no conflict of interest.

References

1. Simões, S. Diffusion Bonding and Brazing of Advanced Materials. *Metals* **2018**, *8*, 959. [CrossRef]
2. Brunatto, S.F.; Klein, A.N.; Muzart, R.J.L. Hollow cathode discharge: Application of a deposition treatment in the iron sintering. *J. Braz. Soc. Mech. Sci. Eng.* **2008**, *30*, 145–151. [CrossRef]
3. Bolotov, G.P.; Bolotov, M.G.; Prybytko, I.O.; Kharchenko, G.K. Diagnosis of plasma glow discharge energy parameters in the processes of treatment small diameter long tubes. In Proceedings of the II International Young Scientists Forum on Applied Physics and Engineering (YSF), Kharkiv, Ukraine, 10–14 October 2016; pp. 116–119. [CrossRef]
4. Metel, A.; Grigoriev, S.; Melnik, Y.; Volosova, M.; Mustafaev, E. Surface Hardening of Machine Parts Using Nitriding and TiN Coating Deposition in Glow Discharge. *Machines* **2020**, *8*, 42. [CrossRef]
5. Petrushynets, L.V.; Falchenko, I.V.; Ustinov, A.I.; Novomlynets, O.O.; Yushchenko, S.M. Vacuum diffusion welding of intermetallic alloy γ TiAl with high-temperature alloy EI437B through nanolayered interlayers. In Proceedings of the 2019 IEEE 2nd Ukraine Conference on Electrical and Computer Engineering (UKRCON), Lviv, Ukraine, 2–6 July 2019; pp. 542–546. [CrossRef]
6. Samouhos, M.; Peppas, A.; Angelopoulos, P.; Taxiarchou, M.; Tsakiridis, P. Optimization of Copper Thermocompression Diffusion Bonding under Vacuum: Microstructural and Mechanical Characteristics. *Metals* **2019**, *9*, 1044. [CrossRef]
7. Borgioli, F.; Galvanetto, E.; Bacci, T. Effects of Surface Modification by Means of Low-Temperature Plasma Nitriding on Wetting and Corrosion Behavior of Austenitic Stainless Steel. *Coatings* **2020**, *10*, 98. [CrossRef]
8. Riccelli, M.G.; Goldoni, M.; Poli, D.; Mozzoni, P.; Cavallo, D.; Corradi, M. Welding Fumes, a Risk Factor for Lung Diseases. *Int. J. Environ. Res. Public Health* **2020**, *17*, 2552. [CrossRef] [PubMed]
9. Bolotov, M.G.; Bolotov, G.P. Elimination of Electric Arc Stabilization in Precision Welding with High-Current Glow Discharge. In Proceedings of the 2019 IEEE 39th International Conference on Electronics and Nanotechnology (ELNANO), Kyiv, Ukraine, 16–18 April 2019; pp. 578–583. [CrossRef]
10. Bolotov, G.P.; Bolotov, M.G.; Stepenko, S.A. The ways of stabilization of high-current glow discharge in welding. In Proceedings of the 2018 IEEE 3rd International Conference on Intelligent Energy and Power Systems (IEPS), Kyiv, Ukraine, 16–18 April 2019; pp. 358–363. [CrossRef]
11. Allis, W.P. Review of glow discharge instabilities. *Physica B+C* **1976**, *82*, 43–51. [CrossRef]
12. Smirnov, S.A.; Baranov, G.A. Gas dynamics and thermal-ionization instability of the cathode region of a glow discharge. Part, I. *Tech. Phys.* **2001**, *46*, 815–824. [CrossRef]
13. Baranov, G.A.; Smirnov, S.A. Nonstationary gas dynamics of the area of a cathodic voltage drop. In Proceedings of the XII International Symposium on Gas. Flow and Chemical Lasers and High-Power Laser Conference, St. Petersburg, Russia, 22 December 1998; p. 820.
14. Bolotov, M.G.; Bolotov, G.P. Criterial Definition of the Limits of Glow Discharge Energy Stability. In Proceedings of the2019 IEEE 2nd Ukraine Conference on Electrical and Computer Engineering (Ukrcon), Lviv, Ukraine, 2–6 July 2019; pp. 497–501. [CrossRef]
15. Megalingam, M.; Sarma, B. Occurrence of ionization instability associated with plasma bubble in glow discharge magnetized plasma. *Plasma Sci. Technol.* **2019**, *21*, 1–22. [CrossRef]
16. Myshenkov, V.I. On the stability of the cathode layer and the mechanism of the transition of a glow discharge into an arc. *Teplofiz* **1984**, *1*, 20–25.
17. Krzhizhanovsky, R.E.; Shtern, Z.Y. Thermophysical properties of non-metallic materials (oxides). *Energy* **1973**, 336. Available online: http://elibrary.nuft.edu.ua/library/DocDescription?doc_id=7459? (accessed on 10 February 2021).
18. Nowak, J.W.; Siemek, K.; Ochał, K.; Kościelniak, B.; Wierzba, B. Consequences of Different Mechanical Surface Preparation of Ni-Base Alloys during High Temperature Oxidation. *Materials* **2020**, *13*, 3529. [CrossRef] [PubMed]
19. Klochko, N.P.; Klepikova, K.S.; Petrushenko, S.I.; Nikitin, A.V.; Kopach, V.R.; Khrypunova, I.V.; Zhadan, D.O.; Dukarov, S.V.; Lyubov, V.M.; Khrypunova, A.L. Effect of Glow-discharge Hydrogen Plasma Treatment on Zinc Oxide Layers Prepared through Pulsed Electrochemical Deposition and via SILAR Method. *J. Nano Electron. Phys.* **2019**, *11*, 5002. [CrossRef]
20. Plesse, H. Untershungen an Electritenlichtbogen. *Ann. D Phys.* **1935**, *22*, 473. [CrossRef]

21. Zhurbenko, V.G.; Nevrovsky, V.A. Thermal processes at the electrodes of the vacuum gap and the initiation of electrical breakdown. Thermal instability of cathode microprotrusions. *Tech. Phys. J.* **1980**, *12*, 2532–2539.
22. Rayzer, Y.P. *Physics of Gas Discharge*, 3rd ed.; Intellect: Moscow, Russia, 2009; p. 590.
23. Landau, L.D. *Electrodynamics of Continuous Media*; Fizmatgiz: Moscow, Russia, 1959; p. 539.
24. Vygodsky, M.Y. *Handbook of Higher Mathematics*; AST Astrel: Moscow, Russia, 2006; p. 991.
25. Hippler, R.; Pfau, S.; Schmidt, M.; Schoenbach, K. *Low Temperature Plasma Physics: Fundamental Aspects and Applications*; Wiley-VCH: Hoboken, NJ, USA, 2001; p. 530.
26. Bolotov, M.G.; Bolotov, G.P. Calculation of the Glow Discharge's Stability Boundary While Welding. In Proceedings of the 2020 IEEE 40th International Conference on Electronics and Nanotechnology (ELNANO), Kyiv, Ukraine, 22–24 April 2020; pp. 775–779. [CrossRef]
27. Lafferty, J.M. Vacuum Arcs. In *Theory and Application*; John Wiley & Sons: Hoboken, NJ, USA, 1980; pp. 5–14.

Article

Surface Discharge Mechanism on Epoxy Resin in Electronegative Gases and Its Application

Herie Park [1], Dong-Young Lim [2] and Sungwoo Bae [3,*]

1. Division of Electrical and Biomedical Engineering, Hanyang University, 222 Wangsimni-ro, Seongdong-gu, Seoul 04763, Korea; bakery@hanyang.ac.kr
2. Hyundai Technical High School, 122 Mipo-ro, Dong-gu, Ulsan 44032, Korea; dylim@hcu.hs.kr
3. Department of Electrical Engineering, Hanyang University, 222 Wangsimni-ro, Seongdong-gu, Seoul 04763, Korea

* Correspondence: swbae@hanyang.ac.kr; Tel.: +82-2-2220-2309

Received: 26 August 2020; Accepted: 23 September 2020; Published: 24 September 2020

Abstract: This study presents the surface discharge characteristics of insulating gases, including sulfur hexafluoride (SF_6), dry air, and N_2, under a non-uniform field. Surface discharge experiments were conducted, with the gas pressure ranging from 0.1 to 0.6 MPa, on samples of epoxy dielectrics under an AC voltage. The experimental results showed that the surface insulation performance significantly improved in insulating gases possessing electronegative gases, such as SF_6 and dry air. Surface flashover voltages of SF_6 were saturated with an increasing pressure, compared to dry air and N_2. The surface discharge mechanism is proposed to explain the improvement and saturation of dielectric characteristics of the electronegative gas in complex dielectric insulations, as well as its influence on the surface flashover voltage. As an application, an insulation design method is discussed with regards to replacing SF_6 gas in high-voltage power equipment based on the knowledge of the physics behind gas discharge.

Keywords: surface discharge; epoxy resin; electronegative gas; high-voltage power equipment

---

1. Introduction

Sulfur hexafluoride (SF_6) plays an important role as an insulating gas in high-voltage power equipment. More than 80% of SF_6 produced worldwide is supplied for use in high-voltage gas circuit breakers, gas-insulated switchgears, and gas-insulated lines, due to its exceptional physical and chemical properties, including its high dielectric strength, current interruption performance, thermal conductivity, thermal stability, nonflammability, nontoxicity, and non-explosive characteristics [1–4].

Despite the outperformance of SF_6, its use in power equipment is becoming increasingly dangerous owing to its high global warming effect, its long atmospheric lifetime, and the high toxicity of decomposed byproducts that result in many environmental problems. The global warming potential (GWP) of SF_6, relative to CO_2, over a 100-year time horizon is 23,500. This means that the global warming effect of 1 kg of SF_6 gas is equivalent to that of 23.5 tons of CO_2 in the atmosphere. Considering that the global annual emission of SF_6 was 8100 tons in 2012, which is equivalent to annual greenhouse gas emissions of approximately 185 million tons of CO_2, its overall GWP impact is not marginal compared to that of CO_2 [5]. Owing to its high greenhouse effect, the European Union intends to reduce the SF_6 sales of 2014 to one-fifth by 2030 [6]. The California Air Resources Board (CARB) has proposed stricter requirements governing the use of SF_6, including phasing out its usage in gas-insulated equipment and further reducing allowable greenhouse gas (GHG) emissions from such equipment from 2020 onwards [7]. Moreover, the decomposition products are produced by the electrical and thermal decomposition of SF_6 in the presence of other molecules, such as H_2O, SiO_2, N_2, O_2, H_2, and Ar. It should be noted that the products containing fluorine and/or sulfur in SF_6 decomposition, including as S_2F_{10}, SF_4, and hydrofluoric acid (HF), are highly reactive, corrosive, and toxic. These are also potent GHGs [8].

Therefore, to reduce its use, many replacements for SF_6 have been investigated for application in high-voltage power equipment. These investigations have focused on using common gases (CO_2, N_2, and dry air), halogenated gases (CF_3I, hydrofluoroolefins (HFOs), and perfluorinated compounds (PFCs)), and mixture gases as substitutes for SF_6. Their dielectric strength and characteristics have been compared with those of SF_6 [9]. In addition to gas insulation, composite insulation has been applied to SF_6-free gas-insulated switchgear (GIS), instead of simple dielectric insulations by vacuum, gases, or solid dielectrics [10,11]. Composite insulation is a method of using a solid dielectric and an insulating gas together. However, composite insulation comes with problems of deterioration in the insulation performance of the equipment, due to the use of a solid dielectric. The insulation performance deteriorates owing to the formation of a triple junction or a surface discharge. The triple junction induces a high electric field in the region where an electrode is in contact with both the insulating gas and the solid dielectric. Surface discharge is undesired gas breakdown that occurs on a surface at a high voltage, lower than the breakdown voltage of the insulating gas. This breakdown is attributed to the weak point of the dielectrics used and their interactions within the power equipment [10,12,13].

To design a composite insulation for SF_6-free power equipment, it is necessary to investigate the fundamental surface discharge characteristics of the composite insulation and identify its weaknesses. Extensive research has been conducted on the surface discharge mechanism in different materials under different electric fields formed by AC, DC, and impulse voltage [10,14–16]. Béroual et al. [10,14,15] presented the experimental characterization of discharge propagation over insulators such as Bakelite, epoxy, and glass immersed in a single gas or mixture gas, under lightning impulse, DC, and AC voltages. The shape and length of creeping discharge patterns were investigated to understand the capacitive charge effect and electric field influence. Kato et al. [16] conducted experiments under a negative impulse voltage to clarify the surface flashover characteristics based on the existence of surface charges on an alumina insulator in a vacuum. The influence of surface charges on the electric field and the secondary electron emission avalanche were explained. These investigations have succeeded in revealing that the surface charges of insulators influence the surface flashover characteristic.

In order to investigate the occurrence of surface flashover, the effects of the gas medium and its pressure, thickness and dielectric constant of a solid insulator, roughness and shape of the electrode, and applied voltage on gas breakdown have been further explored [3,17–21]. These investigations have reported that surface flashover occurs at the triple junction and in its vicinities in several common and specified factors influencing the surface discharge [19]. Park et al. [3] compared the surface discharge characteristics of different gas media, such as SF_6, N_2/O_2 mixture gases, and imitation air, and explained the important role of the O_2 concentration in the N_2/O_2 mixed gas. Lim et al. [19] conducted a comparative study on SF_6 alternative candidate gases for investigating the surface insulation performance of eco-friendly gas, including N_2/O_2 mixture gas, dry air, and compressed air under AC voltage. Based on the experimental results and the comparative study, the effect of moisture contained in the candidate gases and the correlation among the electric filed intensity, collision-ionization coefficient, electron attachment coefficient, and gap length were discussed. Douar et al. [20] presented the ignition of surface discharges and their propagation governed by repulsion due to electrostatic force and attraction caused by nonlocal photo-ionization.

Before the occurrence of surface flashover, the solid dielectric causes the occurrence of physical phenomena, such as electron emission at a triple junction, surface charge accumulation, photoemission, and a secondary electron emission avalanche (SEEA). These phenomena contribute to electron avalanches and the formation of a conductive channel on the dielectric surface. There are several surface discharge mechanisms, based on electron generation and the high electric field as a physical phenomenon in a vacuum [16,18,22]. However, this physical phenomenon and the detailed progression of surface discharge have not yet been fully understood in compressed gases with various electronegative gases. This means that prior research dealing with the progression and suppression of surface discharge is rather limited, in terms of the detailed physical processes of electron generation and disappearance in compressed gas. Although the authors have studied the effect of the concentration of O_2 on surface discharge characteristics in compressed gas in previous works [3,19], many studies on surface discharge seem to be carried out under vacuum or gases with a pressure of 0.3 MPa or less. This research trend has led to a gap in knowledge about the detailed mechanism by which an electronegative gas among high-pressure gases generates electrons during the surface discharge process. This knowledge gap means that the surface insulation design of SF_6-free high-voltage equipment with high-pressure compressed gas is still a challenging problem. In this paper, we analyze the effect of SF_6 and O_2 in compressed gas on electron generation during the surface discharge progress, and provide useful physical information for the surface insulation design of high-pressure SF_6-free high-voltage equipment.

This study describes the surface discharge characteristics of epoxy resin in compressed gases (i.e., SF_6, dry air, and N_2) under a non-uniform electric field. It aims to understand the solid-gas composite insulation characteristics and analyze the effect of secondary electron attachment emitted from the dielectric surface on the surface insulation performance. The reason for the excellent surface insulation performance of electronegative gases (SF_6 and dry air) is explained in detail in this paper, based on the physics governing the gas discharge and the surface discharge mechanisms.

The remainder of this paper is organized as follows. The experimental setup and experimental procedure are introduced in Section 2. The experimental results of the surface discharge characteristics in compressed gas dielectrics are presented in Section 3. The surface discharge mechanism and the insulation design method of an SF_6-alternative insulating gas, based on the effect of the electronegative gases and electron swarm parameters, are discussed in Section 4. Finally, conclusions are presented in Section 5.

2. Experimental Setup and Method

A parallel needle-plane electrode system enclosed in an experimental gas-insulated switchgear (GIS) filled with different gases was used in this study. The specifications of the electrodes used in this experiment are as follows: The needle electrode (N) had a diameter of 5 mm and 20° angle of a point (θ). The diameter of the plane electrode (P) was 59 mm. Both electrodes were made of stainless steel and were arranged vertical to each other. The solid dielectric used was made of epoxy resin and was in the form of a disc with a diameter (d) of 100 mm and a thickness (t) of 2 mm. The sample of epoxy resin was inserted between the electrodes. The distance between the epoxy dielectric and the needle electrode was zero, and the needle electrode was carefully contacted, without damaging the epoxy surface, since damage to the epoxy surface affects the surface flashover voltage. Figure 1a illustrates a schematic diagram of the experimental electrodes and a solid dielectric arrangement.

The experimental gas-insulated switchgear (GIS) chamber is shown in Figure 1b. This chamber is made of stainless steel of a 20 mm thickness and consists of two layers: An interior part with a diameter of 260 mm, height of 460 mm, and volume of 25 L, and an exterior part with a diameter of 460 mm, height of 500 mm, and volume of 83 L. Two windows were used to observe the inside of the experimental chamber. These windows were made of acryl with a diameter of 110 mm and thickness of 20 mm. The chamber was designed and manufactured to endure a fixed temperature range (−90–100 °C) and pressure up to 1.0 MPa. It was also capable of withstanding an AC voltage of 300 kV. A pressure gauge was installed to measure the pressure inside the chamber. The chamber could be preserved at a pressure of 6.67×10^{-2} Pa by using a vacuum pump (SINKU KIKO Co., Ltd. Miyazaki, Japan, GUD-050A).

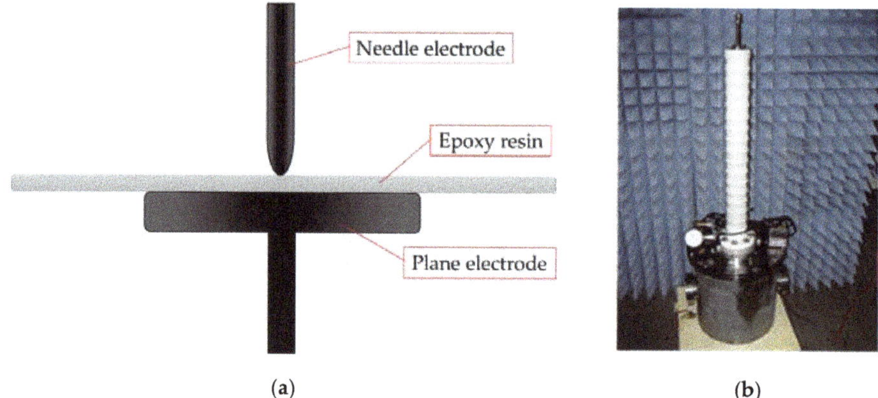

(a) (b)

Figure 1. Experimental setup: (**a**) Scheme of the electrode arrangement and (**b**) the experimental GIS chamber.

The insulation gases used were SF_6, dry air, and N_2. Dry air was produced by a dry air production device. The device had three types of filters that could reduce impurities and lower the dew point of dry air. The final dew point of dry air produced using this device was less than −60 °C. These gases were pressurized from 0.1 to 0.6 MPa, inside the chamber of the device.

The experiments were performed under AC voltage supplied by a high-voltage generator (DY-106-Korea, AC 300 kV/120 mA) with a voltage rising speed of 3.15 kV/s. As an AC power source is actually used as the operating power of GIS, gas-insulated transmission lines (GIL), and gas circuit breaker (GCB), it is very important to analyze the surface discharge characteristics and mechanisms for the device and equipment working under AC power. After ventilation of the chamber to 6.67×10^{-2} Pa, each type of gas was inserted and pressurized from 0.1 to 0.6 MPa, inside the chamber. AC voltage was applied to the electrodes at each pressure level that the gases were subject to. Then, the surface flashover voltages were sequentially measured five times. The average value of this measurement is referred to as the mean surface flashover voltage (V_B) in this study. The surface flashover voltages were measured using a high-voltage probe.

3. Experimental Results

Figure 2 illustrates the surface flashover voltages of SF_6, dry air, and N_2, each at a gas pressure ranging from 0.1 to 0.6 MPa. From this, the effect of pressure on the surface flashover voltages and the breakdown voltages can be analyzed. These voltages increase monotonically with an increasing pressure of SF_6, dry air, and N_2. This monotonous rise is attributed to a shorter mean free path with an increasing gas pressure. The short mean free path limits the acceleration of electrons and reduces the number of electrons with high kinetic energy between the electrodes. The limitation and reduction suppress collision ionization and the SEEA, and such suppression results in an increase in the breakdown voltage and surface flashover voltage. An interesting observation in Figure 2 is the surface discharge characteristic of SF_6. The insulation characteristics of electronegative gases such as SF_6, under a non-uniform electric field, are a clear *N*-shaped breakdown voltage. This *N*-shape rises after the breakdown voltage decreases with an increase in pressure [23,24]. The *N*-characteristic has also been observed in N_2/SF_6 mixed gas containing 0.1% SF_6 [24]. However, the *N* characteristics were not revealed in this study. The corona stabilization effect due to positive space charges leads to *N*-shaped properties. It is presumed that the unobserved *N*-characteristic in this study is due to the weak corona stabilization effect, because of the different experimental conditions employed by other researchers [23,24]. The most evident difference between the references [23,24] and our experiment is the presence of a solid dielectric. The existence of the solid dielectric may strongly influence the distribution of positive ions between electrodes. The mechanism of surface discharge in our electrode system with a solid dielectric would be different from the mechanism of breakdown in an electrode system without any dielectric. The dominant mechanism of surface discharge is the SEEA from the solid dielectric, while that of breakdown manifests as additional avalanches due to positive space charges.

The solid dielectric placed between electrodes plays an important role as an electron source that emits electrons through electron collision and photon irradiation on the surface of the dielectric. Therefore, the dielectric leads the electron generation mechanism, including SEEA and photoemission, while developing surface discharge over the solid dielectric. In contrast, this electron generation mechanism does not occur in the breakdown mechanism under the electrode system without any solid dielectric between electrodes.

Figure 2 shows that the surface flashover voltage measured with a solid dielectric is higher than the breakdown voltage of the electrode system without any solid dielectric. In general, as the solid dielectric forms a triple junction point with a high electric field and electron sources, it is known that the surface flashover voltage is lower than the breakdown voltage. However, our experimental results showed the opposite. This is because the diameter of the solid dielectric is considerably larger than that of the plane electrode in our electrode system.

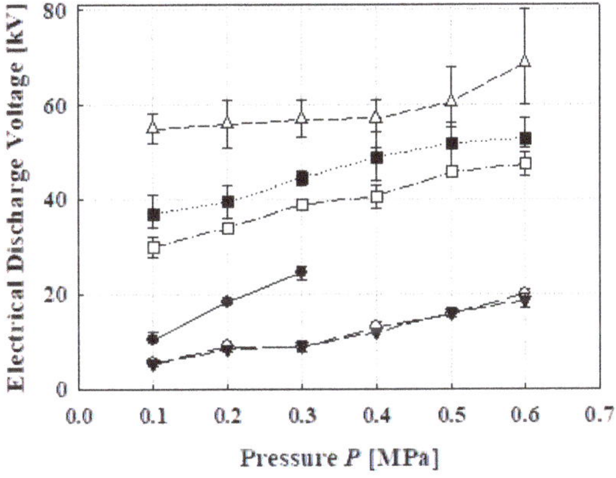

Figure 2. Surface flashover voltage (Δ, SF_6; ■, dry air; and □, N_2) and breakdown voltage (•, SF_6; ○, dry air; and ▼, N_2).

The surface insulation performance and the rate of increase in surface flashover voltages with an increasing pressure are noticeably different for electronegative gases (i.e., SF_6 and dry air) and the non-electronegative gas (i.e., N_2). The surface insulation performance of SF_6 and dry air was better than that of N_2 in the gas pressure range of this experiment. SF_6, being an electronegative gas, demonstrated a higher surface insulation performance than that of dry air. The excellent surface insulation performance of electronegative gases (SF_6 and dry air) is due to the electron attachment mechanism. Although both SF_6 and dry air are electronegative gases, the difference in the surface insulation performance between these gases was possibly due to the electron attachment cross section, which represents the electron attachment capability of gases. Saturation of the surface flashover voltages with an increasing pressure was observed in SF_6—evidently more than in dry air and N_2—within a gas pressure range of 0.1 to 0.4 MPa.

The rate of increase of the surface flashover voltage with an increasing pressure is shown in Figure 3. The insulating gas containing N_2 was insensitive to saturation in the pressure range. This saturation is presumed to be due to the electron detachment of negative ions in the electronegative gases (i.e., SF_6 and O_2) during surface discharge development. As N_2 cannot attach electrons, negative ions are not formed in N_2. This means that no electron detachment mechanism occurs in N_2. A detailed interpretation of the surface insulation performance and surface flashover voltage saturation will be explained in Section 4 based on surface discharge mechanisms.

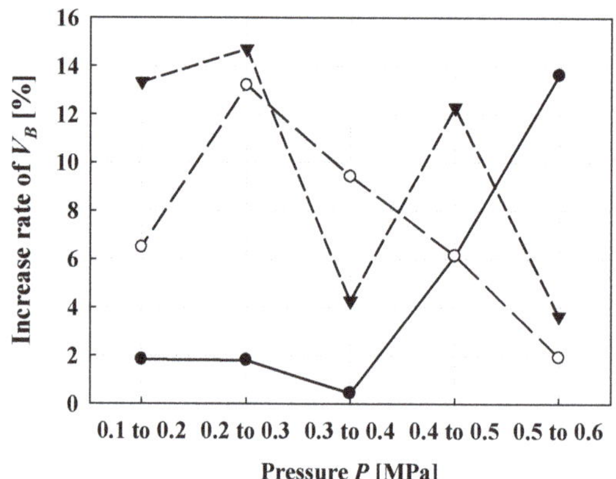

Figure 3. Rate of increase in the surface flashover voltages (•, SF_6; ○, dry air; and ▼, N_2).

4. Discussion

4.1. Surface Discharge Mechanism

This section describes the surface discharge mechanism in detail, explaining the experimental results presented in Section 3. The main outcomes were an improved surface insulation performance of insulating gases containing electronegative species and noticeable saturation of the surface insulation performance of SF_6 with an increasing pressure. These two results seem to be inherently related to the mechanisms of electron attachment and electron detachment, respectively, during the progression of surface discharge. The electron attachment is a mechanism by which SF_6 and O_2 trap electrons moving to the positive electrode to form negative ions between two electrodes. This reduces the number of electrons that cause collision ionization and the SEEA on the dielectric surface, thereby resulting in a higher surface insulation performance from the inclusion of an electronegative gas. However, the negative ions generated by electron attachment emit electrons from collisions among electrons, positive ions, and negative ions. As this electron detachment adds electrons between the electrodes, it may limit the rise of the surface flashover voltage.

Surface discharge is developed by an initial electron, the SEEA, a surface streamer, and conductive channel formation. The formation of an initial electron depends on the polarity of the applied voltage. When the applied voltage is positive, the initial electron is generated by the electron detachment of a negative ion around the needle electrode or electron emission from the epoxy surface. Examples of electron detachment from SF_6^- ions can be presented as follows under an appropriate condition for each case [25]:

Collisional detachment: $SF_6^- + SF_6 \rightarrow SF_6 + SF_6 + e,$

Photodetachment: $SF_6^- + hv \rightarrow SF_6 + e,$

Thermally-induced detachment: $SF_6^- + heat \rightarrow SF_6 + e.$

When the applied voltage is negative, an initial electron is generated by field emission from the cathode at the triple junction, where the electrode, solid dielectric, and insulating gas make contact. This electron collides with the epoxy surface. Then, the generated initial electron contributes to the emission of secondary electrons from the surface. The secondary electrons again lead to an SEEA at the epoxy surface and collision ionization in the gas. During the process of collision ionization, photons emitted from the excited molecules are irradiated to both the epoxy surface and other molecules.

This irradiation results in photoemission and photoionization and can supply additional electrons in the discharge space between the electrodes. The electrons generated by field emission, the SEEA, ionization collision, photoemission, and photoionization can attach themselves to an electronegative gas.

During the development of surface discharge, negative ions can form from the electron attachment process. However, these ions can also disappear due to the electron detachment mechanism during the same surface discharge development. As the surface discharge develops with an increase in the applied voltage, the SEEA and photoemission can become more prominent. Owing to the active SEEA and photoemission processes, a surface streamer develops along the epoxy surface. The head of the streamer advances based on what the electron source is (photoionization/photoemission), and the surface streamer intersects between the electrodes. When a conductive channel is formed between the electrodes due to the surface streamer and the presence of many electrons, surface flashover finally occurs.

The most evident difference between an electronegative gas and a non-electronegative gas is the occurrence of electron attachment or detachment during the process of surface discharge development. The authors believe that these processes are the key determining factors that explain the experimental results. Electron attachment is the only process that removes electrons in the discharge space between electrodes in the surface discharge mechanism. The removal of electrons suppresses the formation of the streamer and conductive channels. This suppression is the reason behind the high surface insulation performance of an electronegative gas. Therefore, the surface insulation performance of the insulating gas containing electronegative gases SF_6 and O_2 is superior to that of N_2.

Although SF_6 and dry air are both electronegative gases, the disparity in the surface insulation performance between these gases can be explained by the difference in the electron attachment cross section, which determines the characteristics of the gases. Since the surface discharge grows on the surface of the epoxy dielectric, the capturing of electrons emitted from the surface appears to contribute to a significant improvement in the surface insulation performance. Electrons are generated from the epoxy surface during surface discharge by the SEEA and photoemission processes. The energy range of the electrons emitted from the surface by these processes was assumed to be 3.0–4.5 eV [16,26]. Figure 4 shows the electron attachment cross sections of SF_6 and O_2 in the electron energy range. Evidently, the cross section of SF_6 is considerably larger than that of O_2. From this figure, it can be assumed that the ability of secondary electrons, emitted from the epoxy surface, to attach themselves is significantly better in SF_6 than in O_2.

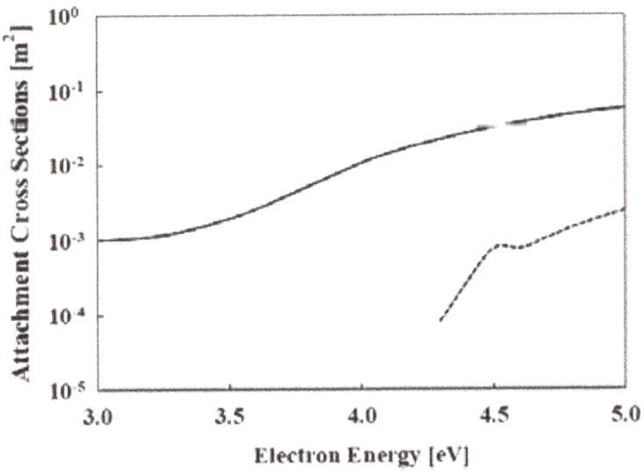

Figure 4. Attachment cross sections of gases (—, SF_6 and ---, O_2).

The progress of the streamer and the formation of a conductive channel on the epoxy surface can be suppressed by the capture (attachment) of the secondary electrons due to the large electron attachment cross sections. This could be the reason why SF_6, which has an excellent electron attachment ability in the electron energy range of 3.0–4.5 eV, exhibits a superior surface insulation performance compared to dry air [27].

In addition, the noticeable saturation of surface flashover voltages with an increasing pressure, in the pressure range of 0.1 to 0.4 MPa in SF_6, may be related to electron detachment. During surface discharge development, negative ions collide with electrons, positive ions, and other negative ions in the medium. Electrons attached to electronegative gas are then separated and detached from the negative ions by these collisions. The separated electrons are imbibed into the SEEA or collision ionization, helping the formation of a conductive path along the epoxy surface between the electrodes. In other words, a large number of negative ions can improve the surface insulation performance, and at the same time, can function as electron sources [28]. The electron detachment of negative ions mainly occurs from collisions with ions rather than high electric fields [29]. Since the mean free path becomes shorter as the pressure increases, it can be assumed that the collision of positive and negative ions also occurs more frequently with an increasing pressure. Frequent collisions activate electron detachment, causing a noticeable saturation of the surface insulation performance with an increasing pressure. The rise of the surface flashover voltage of SF_6 at 0.6 MPa may be related to a high-pressure surface discharge mechanism, such as leader discharge.

4.2. Insulation Design Method of an Insulating Gas Alternative to SF_6 Based on Electron Swarm Parameters

An insulation design method that can ensure a good insulation performance in high-voltage power apparatus such as GIS can be proposed from our experimental results, derived from Figure 2. The surface insulation performance can be enhanced by the inclusion of electronegative gases and an increase in the gas pressure. Therefore, these factors must be considered during insulation design. However, very rarely is apparatus designed with a gas pressure exceeding 1.0 MPa [23]; therefore, it is desirable to incorporate the addition of electronegative gases into the insulation design method.

An insulation design using a mixture of electronegative gases should be carried out upon analyzing the electron swarm parameters of the gases. These parameters are the electron attachment cross section and electron scattering cross section. The electron attachment cross section represents the ability of an electronegative gas to capture electrons. This ability depends on the energy of the electrons. There are certain energy ranges in which electronegative gases can effectively and easily capture electrons. In the discharge space, electrons accelerated by the applied electric field are distributed with various values of kinetic energy. The improvement in the insulation performance of the gas is attributed to the easy and effective capture of these electrons. Therefore, there is a need for a method of selecting an electronegative gas with a large electron attachment cross section spanning a wide range of electron energies. The lower the kinetic energy of the electrons, the easier the electron attachment is; thus, for insulation design, it is preferable to select an electronegative gas possessing a wide electron attachment cross section in a low electron energy range.

Meanwhile, the electron scattering cross section represents the ability to reduce the kinetic energy of electrons, which are accelerated by the applied electric field. This reduction of kinetic energy effectively activates the electron attachment of electronegative gases. Therefore, electron scattering results in high-energy electrons with the sufficient ionization capacity becoming inactive, consequently resulting in effective electron attachment.

Since the electron attachment process is more important than the ionization process in the prevention of a breakdown, only the addition of an electronegative gas is usually conducted in the insulation design method. However, for producing a higher insulation performance, the electron swarm parameters cannot be overlooked. Based on these two parameters (the electron attachment cross section and electron scattering cross section), the mixture gas that is more suitable for insulation design should include a gas with electron attachment abilities and the ability to suppress the electron behavior. It is preferable to include a mixture of two or more auxiliary gases with different electron attachment cross section peaks, depending on the electron energy, because one electronegative gas cannot capture all electrons in a considerably wide electron energy region. Meanwhile, the mixing of gases with an electron scattering capability that reduces the electron energy in the electron energy region corresponding to the peak of the electron attachment cross section should be recommended.

Based on these electron swarm parameters, the experimental results reported that the 2.5% positive lightning impulse breakdown voltage of SF_6/air and SF_6/CO_2 in 0.05 to 0.45 MPa under a non-uniform electric field with a gap of 10 mm is higher than that of pure SF_6 [30]. Kojima et al. [31] studied the insulation properties of a mixed gas containing two or more electronegative gases based on the electron attachment cross section, depending on the electron energy.

In addition to selecting the insulating gas, engineering techniques can be applied to an insulation design method, and these techniques should be integrated into high-voltage apparatus with a selection method of insulating gas that considers the electron swarm parameters. The engineering techniques are improvements in electrodes and solid dielectrics, details of which can be found in references [11,32–36]. The theoretical principles governing the engineering techniques are the suppression of electron generation and relaxation of the concentrated electric field. Since suppression and relaxation are the direct factors influencing gas breakdown, a good insulation performance can also be ensured with these engineering techniques. Therefore, when insulation gas selection and engineering insulation techniques are simultaneously applied to high-voltage apparatus, the surface insulation performance of the apparatus can be improved, owing to a synergistic effect that imbibes physical and engineering elements.

Therefore, the selection of insulating gas based on the physics of gas discharge (i.e., electron swarm parameters) is more important than engineering insulating design techniques. This is because engineering insulation design techniques are based on physical knowledge (i.e., physical mechanisms and physical analysis), while the insulation performance of the high-voltage apparatus is significantly dependent on the insulating gas. To date, single or mixed insulating gases that can completely replace SF_6 in terms of insulation, arc quenching, and liquefaction have not yet been reported in the fields of applied physics and high-voltage engineering. The authors believe that a new insulating gas as an alternative to SF_6 can be discovered or developed, based on the physics-based knowledge of gas discharge.

5. Conclusions

This study investigated the surface discharge characteristics of epoxy resin in compressed gases, namely SF_6, dry air, and N_2, in a non-uniform electric field. Among the different gases, SF_6 and dry air, which are electronegative gases, revealed better insulation performances compared with N_2, owing to their electron attachment ability. The effect of electronegative gases on surface flashover voltages, varying with the pressure in these compressed gases, was analyzed in detail through the processes of electron attachment and electron detachment. Electron attachment induced an increase in the surface flashover voltage of the electronegative gas. Electron detachment seemed to result in saturation of the voltages as the pressure increased. A surface discharge mechanism that includes electron attachment, electron detachment, and secondary electron emission (SEEA and photoemission) was proposed for an analysis of the surface flashover voltages, which varied with the gas pressure. From this mechanism, it was found that the physical phenomena governing the behavior of electrons, including electron attachment, electron detachment, and electron scattering, are related to an improvement in the dielectric strength of the mixed gas. Based on these findings, techniques for improving the dielectric strength of mixed gases were discussed from a physics and engineering point of view. The discussions and knowledge of the physics behind gas discharge suggest the possibility of a new insulating gas to replace SF_6. To discover and develop the new insulating gas, theoretical and experimental attempts to understand the effects of varying the pressure on electron attachment and electron detachment are required. In addition to such attempts, the insulation characteristics of gases and mixed gases and their discharge mechanisms under DC and AC voltages must specifically be identified. This study contributes to explorations attempting to expand the knowledge of physics that governs gas discharge. The physical information obtained from the results of this study can be used to secure improved surface insulation performances by using the technique of mixing insulation gas in the surface insulation design of SF_6-free high-voltage equipment.

Author Contributions: Conceptualization, H.P., D.-Y.L. and S.B.; Investigation, H.P., D.-Y.L. and S.B.; Supervision, S.B.; Writing—original draft, H.P., D.-Y.L. and S.B.; Writing—review & editing, H.P., D.-Y.L. and S.B. All of the authors were involved in the preparation of the manuscript. All authors have read and agreed to the published version of the manuscript.

Funding: This research was funded by the Basic Science Research Program through the National Research Foundation of Korea (NRF) funded by the Ministry of Education, grant number 2017R1D1A3B03035693. This work was also supported by the Korea Institute of Energy Technology Evaluation and Planning (KETEP) grant funded by the Korea government (MOTIE) (No. 20192010107050).

Conflicts of Interest: The authors declare no conflict of interest.

References

1. Pan, B.; Wang, G.; Shi, H.; Shen, J.; Ji, H.-K.; Kil, G.-S. Green Gas for Grid as an Eco-Friendly AlternativeInsulation Gas to SF_6: A Review. *Appl. Sci.* **2020**, *10*, 2526. [CrossRef]
2. Xiao, S.; Zhang, X.; Tang, J.; Liu, S. A review on SF_6 substitute gases and research status of CF_3I gases. *Energy Rep.* **2018**, *4*, 486–496. [CrossRef]
3. Park, H.; Lim, D.-Y.; Bae, S. Partial discharge and surface flashover characteristics with O_2 content in N_2/O_2 mixed gas under a non-uniform field. *IEEE Trans. Dielectr. Electr. Insul.* **2018**, *25*, 1403–1412. [CrossRef]
4. Beroual, A.; Haddad, M.A. Recent Advances in the Quest for a New Insulation Gas with a Low Impact on the Environment to Replace Sulfur Hexafluoride (SF_6) Gas in High-Voltage Power Network Applications. *Energies* **2017**, *10*, 1216. [CrossRef]
5. Dunse, B.L.; Fraser, P.J.; Krummel, P.B.; Steele, L.P.; Derek, N. Australian HFC, PFC, Sulfur Hexafluoride and Sulfuryl Fluoride Emissions. Report Prepared for Australian Government Department of the Environment, by the Collaboration for Australian Weather and Climate Research, CSIRO Oceans and Atmosphere Flagship 2015, Aspendale, Australia, iv, 28. Available online: https://www.environment.gov.au/system/files/resources/71fad794-d98d-4014-95e6-53e9ce6493bf/files/australian-hfc-pfc-emissions-2015.pdf (accessed on 23 September 2020).
6. Mota-Babiloni, A.; Navarro-Esbrí, J.; Barragán-Cervera, Á.; Molés, F.; Peris, B. Analysis based on EU Regulation No 517/2014 of new HFC/HFO mixtures as alternatives of high GWP refrigerants in refrigeration and HVAC systems. *Int. J. Refrig.* **2015**, *52*, 21–31. [CrossRef]
7. California Air Resource Board, E.T.a.D.G.G.E. Electricity Transmission and Distribution Greenhouse Gas Emissions. Available online: https://ww2.arb.ca.gov/our-work/programs/elec-tandd (accessed on 23 September 2020).
8. Tsai, W.-T. The decomposition products of sulfur hexafluoride (SF_6): Reviews of environmental and health risk analysis. *J. Fluor. Chem.* **2007**, *128*, 1345–1352. [CrossRef]
9. Li, X.; Zhao, H.; Wu, J.; Jia, S. Analysis of the insulation characteristics of CF_3I mixtures with CF_4, CO_2, N_2, O_2 and air. *J. Phys. D Appl. Phys.* **2013**, *46*, 345203. [CrossRef]
10. Beroual, A.; Coulibaly, M.; Aitken, O.; Girodet, A. Investigation on creeping discharges propagating over epoxy resin and glass insulators in the presence of different gases and mixtures. *Eur. Phys. J. Appl. Phys.* **2011**, *56*, 30802. [CrossRef]
11. Goshima, H.; Okabe, S.; Ueda, T.; Morii, H.; Yamachi, N.; Takahata, K.; Hikita, M. Fundamental insulation characteristics of high-pressure CO_2 gas for gas-insulated power equipment—Effect of coating conductor on insulation performance and effect of decomposition products on creeping insulation of spacer. *IEEE Trans. Dielectr. Electr. Insul.* **2008**, *15*, 1023–1030. [CrossRef]
12. Li, Z.; Okamoto, K.; Ohki, Y.; Tanaka, T. The role of nano and micro particles on partial discharge and breakdown strength in epoxy composites. *IEEE Trans. Dielectr. Electr. Insul.* **2011**, *18*, 675–681. [CrossRef]
13. Li, S.; Huang, Y.; Min, D.; Qu, G.; Niu, H.; Li, Z.; Wang, W.; Li, J.; Liu, W. Synergic effect of adsorbed gas and charging on surface flashover. *Sci. Rep.* **2019**, *9*, 5464. [CrossRef] [PubMed]
14. Sadaoui, F.; Beroual, A. AC creeping discharges propagating over solid–gas interfaces. *IET Sci. Meas. Technol.* **2014**, *8*, 595–600. [CrossRef]
15. Sadaoui, F.; Beroual, A. DC creeping discharges over insulating surfaces in different gases and mixtures. *IEEE Trans. Dielectr. Electr. Insul.* **2014**, *21*, 2088–2094. [CrossRef]

16. Kato, K.; Kato, H.; Ishida, T.; Okubo, H.; Tsuchiya, K. Influence of surface charges on impulse flashover characteristics of alumina dielectrics in vacuum. *IEEE Trans. Dielectr. Electr. Insul.* **2009**, *16*, 1710–1716. [CrossRef]
17. Sudarshan, T.S.; Dougal, R.A. Mechanisms of Surface Flashover Along Solid Dielectrics in Compressed Gases: A Review. *IEEE Trans. Electr. Insul.* **1986**, 727–746. [CrossRef]
18. Li, C.R.; Sudarshan, T.S. Dielectric surface preflashover processes in vacuum. *J. Appl. Phys.* **1994**, *76*, 3313–3320. [CrossRef]
19. Lim, D.-Y.; Bae, S. Study on oxygen/nitrogen gas mixtures for the surface insulation performance in gas insulated switchgear. *IEEE Trans. Dielectr. Electr. Insul.* **2015**, *22*, 1567–1576. [CrossRef]
20. Douar, M.A.; Beroual, A.; Souche, X. Ignition and advancement of surface discharges at atmospheric air under positive lightning impulse voltage depending on perpendicular electric stress and solid dielectrics: Modelling of the propagating phenomenology. *Eur. Phys. J. Appl. Phys.* **2018**, *82*, 20801. [CrossRef]
21. Yan, X.; Zheng, Y.; Gao, K.; Xu, X.; Wang, W.; He, J. The Criterion of Conductor Surface Roughness in Environment-friendly C_4F_7N/CO_2 Gas Mixture. In Proceedings of the 2019 IEEE Sustainable Power and Energy Conference (iSPEC); Institute of Electrical and Electronics Engineers (IEEE), Bejing, China, 21–23 November 2019; pp. 377–381.
22. Nakano, Y.; Kojima, H.; Hayakawa, N.; Tsuchiya, K.; Okubo, H. Pre-discharge and flashover characteristics of impulse surface discharge in vacuum. *IEEE Trans. Dielectr. Electr. Insul.* **2014**, *21*, 403–410. [CrossRef]
23. Hikita, M.; Ohtsuka, S.; Okabe, S.; Kaneko, S. Insulation characteristics of gas mixtures including perfluorocarbon gas. *IEEE Trans. Dielectr. Electr. Insul.* **2008**, *15*, 1015–1022. [CrossRef]
24. Yializis, A.; Malik, N.; Qureshi, A.; Kuffel, E. Impulse Breakdown and Corona Characteristics for Rod-Plane Gaps in Mixtures of SF_6 and Nitrogen with Less Than 1% of SF_6 Content. *IEEE Trans. Power Appar. Syst.* **1979**, *5*, 1832–1840. [CrossRef]
25. Christophorou, L.G.; Olthoff, J.K. Electron Interactions with SF_6. *J. Phys. Chem. Ref. Data* **2000**, *29*, 267–330. [CrossRef]
26. Hosokawa, T.; Kaneda, T.; Takahashi, T.; Yamamoto, T.; Morita, T.; Sekiya, Y. DC breakdown characteristics in the gap with thin dielectric sheet in air. *IEEE Trans. Dielectr. Electr. Insul.* **2011**, *18*, 822–832. [CrossRef]
27. Raju, G.G. *Gaseous Electronics*; CRC Press: Boca Raton, FL, USA, 2012.
28. Teich, T.H. *Detachment of Electrons from Negative Ions in Electrical Discharges*; Springer Science and Business Media LLC: Boston, MA, USA, 1991; pp. 215–229.
29. Christophorou, L.; Pinnaduwage, L. Basic Physics of Gaseous Dielectrics. *IEEE Trans. Electr. Insul.* **1990**, *25*, 55–74. [CrossRef]
30. Qiu, X.Q.; Chalmers, I.D.; Coventry, P. A study of alternative insulating gases to SF_6. *J. Phys. D Appl. Phys.* **1999**, *32*, 2918–2922. [CrossRef]
31. Kojima, H.; Kinoshita, O.; Hayakawa, N.; Endo, F.; Okubo, H.; Yoshida, M.; Ogawa, T. Breakdown Characteristics of N_2O Gas Mixtures for Quasiuniform Electric Field under Lightning Impulse Voltage. *IEEE Trans. Dielectr. Electr. Insul.* **2007**, *14*, 1492–1497. [CrossRef]
32. Harris, J.R.; Kendig, M.; Poole, B.; Sanders, D.M.; Caporaso, G. Electrical strength of multilayer vacuum insulators. *Appl. Phys. Lett.* **2008**, *93*, 241502. [CrossRef]
33. Kurimoto, M.; Kato, K.; Hanai, M.; Hoshina, Y.; Takei, M.; Okubo, H. Application of functionally graded material for reducing electric field on electrode and spacer interface. *IEEE Trans. Dielectr. Electr. Insul.* **2010**, *17*, 256–263. [CrossRef]
34. Okubo, H. Enhancement of electrical insulation performance in power equipment based on dielectric material properties. *IEEE Trans. Dielectr. Electr. Insul.* **2012**, *19*, 733–754. [CrossRef]
35. Rokunohe, T.; Yagihashi, Y.; Endo, F.; Oomori, T. Fundamental insulation characteristics of air; N_2, CO_2, N_2/O_2, and SF_6/N_2 mixed gases. *Electr. Eng. Jpn.* **2006**, *155*, 9–17. [CrossRef]
36. Rokunohe, T.; Yagihashi, Y.; Aoyagi, K.; Oomori, T.; Endo, F. Development of SF_6-Free 72.5 kV GIS. *IEEE Trans. Power Deliv.* **2007**, *22*, 1869–1876. [CrossRef]

MDPI
St. Alban-Anlage 66
4052 Basel
Switzerland
Tel. +41 61 683 77 34
Fax +41 61 302 89 18
www.mdpi.com

Applied Sciences Editorial Office
E-mail: applsci@mdpi.com
www.mdpi.com/journal/applsci

www.ingramcontent.com/pod-product-compliance
Lightning Source LLC
LaVergne TN
LVHW070613170726
843515LV00004B/64